High-Entropy Materials

Research in the field of high-entropy materials is advancing rapidly. *High-Entropy Materials: Advances and Applications* focuses on materials discovered using the high-entropy alloys (HEA) strategy. It discusses various types of high-entropy materials, such as face-centered cubic (FCC) and body-centered cubic (BCC) HEAs, films and coatings, fibers, and powders and hard-cemented carbides, along with current research status and applications:

- Describes, compositions and processing of high-entropy materials.
- Summarizes industrially valuable alloys found in high-entropy materials that hold promise for promotion and application.
- Explains how high-entropy materials can be used in many fields and can outperform traditional materials.

This book is aimed at researchers, advanced students, and academics in materials science and engineering and related disciplines.

Emerging Materials and Technologies

Series Editor: Boris I. Kharissov

The *Emerging Materials and Technologies* series is devoted to highlighting publications centered on emerging advanced materials and novel technologies. Attention is paid to those newly discovered or applied materials with potential to solve pressing societal problems and improve quality of life, corresponding to environmental protection, medicine, communications, energy, transportation, advanced manufacturing, and related areas.

The series takes into account that, under present strong demands for energy, material, and cost savings, as well as heavy contamination problems and worldwide pandemic conditions, the area of emerging materials and related scalable technologies is a highly interdisciplinary field, with the need for researchers, professionals, and academics across the spectrum of engineering and technological disciplines. The main objective of this book series is to attract more attention to these materials and technologies and invite conversation among the international R&D community.

Functional Biomaterials
Advances in Design and Biomedical Applications
Anuj Kumar, Durgalakshmi Dhinasekaran, Irina Savina, and Sung Soo Han

Smart Nanomaterials
Imalka Munaweera and M. L. Chamalki Madhusha

Nanocosmetics
Drug Delivery Approaches, Applications and Regulatory Aspects
Edited by: Prashant Kesharwani and Sunil Kumar Dubey

Sustainability of Green and Eco-friendly Composites
Edited by Sumit Gupta, Vijay Chaudhary, and Pallav Gupta

For more information about this series, please visit: www.routledge.com/
Emerging-Materials-and-Technologies/book-series/CRCEMT

High-Entropy Materials
Advances and Applications

Yong Zhang

CRC Press
Taylor & Francis Group
Boca Raton London New York

CRC Press is an imprint of the
Taylor & Francis Group, an **informa** business

First edition published 2024
by CRC Press
6000 Broken Sound Parkway NW, Suite 300, Boca Raton, FL 33487–2742

and by CRC Press
4 Park Square, Milton Park, Abingdon, Oxon, OX14 4RN

CRC Press is an imprint of Taylor & Francis Group, LLC

ISBN: 978-1-032-32391-6 (hbk)
ISBN: 978-1-032-33520-9 (pbk)
ISBN: 978-1-003-31998-6 (ebk)

DOI: 10.1201/9781003319986

Typeset in Times LT Std
by Apex CoVantage, LLC

Contents

Preface

Materials science is mainly focused on the relationship between microstructures and properties. The properties of a material are determined by the combination of components, atomic arrangement, and internal microstructures. Entropy, as a parameter in thermodynamics and statistics, represents the spontaneous evolution direction, arrangement of the components in the space, and disorder or chaos, which also indicates the structures. According to Shannon, entropy is inversely proportional to the amount of information, which actually reflects information as well; actually, structures represent the information. High-entropy alloys (HEAs) are developed new materials from the point of view of the chemical disorder.

The unique properties are permanent pursuits for human beings. The unique properties are definitely related to the special microstructures of the materials. To realize the special microstructures, two typical strategies are usually used. One is to adjust the compositions; here, we are using the high-entropy compositions, which may make the microstructures special and complex (e.g., adjustable lattice distortion; face-centered cubic (FCC) to body-centered cubic (BCC) with changed Al content; lower stacking-fault energy, making the twinning and phase change easy during the plastic deformation at cryogenic temperature and high strain rate loading; and the high-entropy composition also makes the dislocation moving peculiar). Another is to use techniques such as fish-bone and dendrite morphology obtained by using the Bridgman solidification technique; bionic bamboo fiber by using cold-drawing methods; and bamboo-joint-structured Cu alloy fiber with shape memory effects and superelastic strain limits prepared by glass-coating through the Taylor technique. This is breaking through the limitation of material primitives and breaking the performance limits of traditional materials, such as overcoming the strength-ductility trade-off. Professor Zhang's research group prepared remarkable toughening high-entropy alloy wire with a bionic bamboo-fiber-heterogeneous (BFH) structure. They have successfully introduced a bionic BFH structure into an $AlCoCrFeNi_{2.1}$ eutectic high-entropy alloy by the cold-drawing process. The BFH wires possess an extraordinary synergism of strength and ductility.

The first chapter of this book is written by Professor Zhang, which mainly discusses the origination of high-entropy alloys and high-entropy materials; Changwei Li mainly introduces the FCC high-entropy alloys of GS101, GS102, and CoCrFeNi series in Chapter 2. In the third chapter, Fangfei Liu introduces BCC HEAs from production methods, mechanical behavior, strengthening, special performance, and application aspects. Shichao Zhou discusses the microstructures and properties of duplex-phase HEAs, the irradiation behavior of duplex-phase HEAs, and eutectic HEAs in Chapter 4. Chapter 5, by Xuehui Yan, introduces the processing of high-entropy films and coating (e.g., sputtering, laser cladding), the properties of the high-entropy films, and the compositional gradient films. Ruixuan Li summarizes the preparation methods and properties of HEA fibers and gives an outlook on their future development in Chapter 6. Chapter 7, written by Yuxin Wen, discusses the processing of powder technology. This chapter includes mechanical alloying by ball milling, EIGA, and VIGA; the processing of high-entropy alloys, GS201 ($AlCoCrFeNiTi_{0.2}$),

GS301 ($AlCo_{0.4}CrFeNi_{2.7}$), and cemented hard carbide alloys. In Chapter 8, Yaqi Wu analyzes the number of papers about high-entropy ceramics published in recent years and the proportion of each type of high-entropy ceramic. In Chapter 9, Yuanying Yue first introduces the high-entropy polymer (HEP); then the high-entropy concept is extended to HEPs. Finally, high-entropy composition materials are introduced. Xinfang Song discusses future trends and applications of high-entropy materials to improve human life and well-being in Chapter 10.

This book summarizes the discovery and recent development of high-entropy materials, which also gives a brief introduction to various high-entropy alloy materials. This book provides an introduction to the reader in order to conduct more in-depth and systematic research in the future.

About the Author

Yong Zhang graduated from Yanshan University, Qinghuang Island, Hebei Province, with a bachelor's degree in 1991, and obtained a master's degree and PhD from the University of Science and Technology Beijing (USTB) in 1994 and 1998, respectively. Then he joined the Institute of Physics (IOP), Chinese Academy of Science (CAS), and worked as Postdoctoral Fellow. In 2000, he joined the Singapore-Massachusetts Institute of Technology (MIT) Alliance (SMA) and the National University of Singapore (NUS), and worked as Research Fellow in the program of Advanced Materials for Micro & Nano-Systems (AMM&NS). In 2004, he was promoted to Full Professor in 2004 at the USTB. In 2005, he was awarded the New Century Excellent Talent (NCET) by the Ministry of Education of China. He was Senior Visiting Scholar at the University of Tennessee, Knoxville, USA.

He prepared the bionic dendrite and fish-bone structure of amorphous and high-entropy composites by using the Bridgman-solidification technique. He prepared the bionic bamboo-fiber structure high-entropy wires by using cold-drawing methods. He prepared the glass-coated Cu alloy fibers with shape-memory effect and super-elastic strain limits with bamboo-joint structures. He prepared the lotus-shaped amorphous porous alloys by using the etching technique. He prepared the first single-crystal high-entropy alloys, $Al_{0.3}CoCrFeNi$ and $Al_{0.3}CoCrFeNi_2$. The invention of trace rare earth elements can improve the glass-forming ability (GFA) of amorphous alloys, which has been widely used in academia and industries. The first body-centered cubic (BBC) alloy with high strength and high entropy was synthesized. The ratio parameter w of information entropy and mixing enthalpy was put forward to evaluate the disorder of materials, which has been proven by a large number of documents to be effective in predicting the formation of random solid-solution and amorphous phases.

High-entropy alloy fiber and high-entropy alloy photo-thermal selective films have been successfully studied. Professor Zhang participated in publishing the monographs *Amorphous and High-Entropy Alloys* (China Science Press, 2010); *Advanced High-Entropy Alloys Technology* (Chemical Industry Press, 2018); *High-Entropy Materials, A Brief Introduction* (Springer Berlin Heidelberg, 2019); *Magnetic Sensors-Development Trends and Applications* (Intechopen, 2017); *Stainless Steels and Alloys* (Intechopen, 2019); *Engineering Steels and High-Entropy Alloys* (Intechopen, 2020); *High-Entropy Alloys, Fundamental and Applications* (Springer, 2018); *High-Entropy Alloys, Innovations, Advances, and Applications* (CRC Press, 2020); and more. He participated in the Ministry of Education Natural Science First Prize and Second Prize, the National Natural Science Second Prize, and the Shanxi Provincial Education Department Natural Science First Prize. He is also Member of the Amorphous Committee of the Metal Society, Fellow of the China Materials Research Society, and Fellow of the Nuclear Materials Society. Professor Zhang participated in organizing the conference on high-entropy alloys and serration behaviors, and served as Chair or Co-chair. He was Flexible Appointed Professor in the North Minzu University. He is also Science and Technology Correspondent of Guangdong Province and Guest Professor at North University of China. He has been selected as one of the thousand talents in Qinghai Province. He has edited the special issues "Serration and Noise Behaviors in Advanced Materials," "Nanostructured High-Entropy Alloys," "The New Advances in High-Entropy Alloys," and more. Professor Zhang devoted himself to studying serration behavior, high-throughput technology, and collective effect in materials science. Professor Zhang has obtained 2 USA patents and 8 China patents, has published more than 200 papers, and has been cited more than 30,000 times. And one paper published in the journal *Progress in Materials Science* has been cited more than 3,000 times. He is highly cited Author by Elsevier. He has published a paper in *Science* and reported in the paper *Nature*. He is Reviewer of many journal papers and books, including *Nature, Science, Nature Reviews Materials*, and *Progress in Materials Science*. He is Section Editor-in-Chief of *Journal of Metals* and *Entropic Alloys and Meta-Metals*, and Associate Editor for *Structural Materials* in the *Frontiers in Materials*. He is also Editorial Committee Member of the journal *Metals and Materials International*. He is Editorial Board Member of the *International Journal of Minerals, Metallurgy, and Materials* (IJMMM), *Metals World, Metals*, and others; and Editorial Advisory Board Member of the journal of *High Entropy Alloys and Materials*.

Professor Zhang's main research interest is the development of new materials in bulk, film, and fiber forms by using high-entropy and Materials Genome Initiative (MGI) strategies to set up the relationship between micro- and nano-scale structures with their properties and performances.

Contributors

Changwei Li
State Key Laboratory for Advanced
Metals and Materials
University of Science and Technology
Beijing (USTB)
Beijing, China

Fangfei Liu
State Key Laboratory for Advanced
Metals and Materials
USTB
Beijing, China

Ruixuan Li
State Key Laboratory for Advanced
Metals and Materials
USTB
Beijing, China

Shichao Zhou
State Key Laboratory for Advanced
Metals and Materials
USTB
Beijing, China

Xinfang Song
State Key Laboratory for Advanced
Metals and Materials
USTB
Beijing, China

Xuehui Yan
State Key Laboratory for Advanced
Metals and Materials
USTB
Beijing, China

Yaqi Wu
State Key Laboratory for Advanced
Metals and Materials
USTB
Beijing, China

Yong Zhang
State Key Laboratory for Advanced
Metals and Materials
USTB
Beijing, China

Yuanying Yue
State Key Laboratory for Advanced
Metals and Materials
USTB
Beijing, China

Yuxin Wen
State Key Laboratory for Advanced
Metals and Materials
USTB
Beijing, China

1 Brief Introduction of High-Entropy Materials

Yong Zhang

1.1 ORIGINATION OF HIGH-ENTROPY ALLOYS

Materials are the basis for civilization's progress in human beings. Materials are usually used based on their properties, and the properties can be mechanical (e.g., strength and ductility), physical (e.g., magnetism, conductivity), or chemical (e.g., corrosion resistance). The performance and applications of materials are related to many properties but may have one or two that are dominant (e.g., structural materials mainly concern the mechanical behaviors), but if the corrosion resistance is excellent, it may allow the materials to be used for a longer time. The properties of materials are related to their compositions, micro- and nano-level structures, and processing technology. The basic principle for materials science is the relationship between structures and properties. The structures of materials can be categorized into grain, grain boundary, domain, dislocations, twinning, and others. The types of structures can be divided into amorphous, crystalline, based order, and disorder. Single crystal (e.g., silicon wafers, diamond); crystalline, steels, and Al alloys; amorphous alloys (e.g., ZrNbTiAlY); and high-entropy alloys, $AlCoCrFeNiTi_{0.2}$. The order and disorder in materials are the basic ideas that might have been proposed by Landau, a famous scientist. The concept of high-entropy alloys was published in 2004 by Professor Yeh JW; he discovered a solid-solution phase without the distinction of solute and solvent elements, which was completely different from the tradition solid solution, also called terminal solid solution, with a solvent and solute; the structure is the same with the solvent, actually, continuous solid solution, just like CuNi (Figure 1.1), which has the same face-centered cubic (FCC) structure.

Before 2004, there existed many scientists who had studied the mixing behavior of many elements and many carbides, nitrides, and borides, and even polymers. Moreover, some scientists have discovered that the high mixing entropy intended to stabilize the random solid-solution phases rather than the intermetallic compound ordered phases. One important alloy is CoCrFeMnNi, which forms FCC solid-solution single phase, also called Cantor alloy because it was discovered by Professor Cantor. Another typical high-entropy alloy is AlCoCrFeNiCu, which was discovered by Professor Yeh; however, because Cu usually shows segregation, Zhang et al. removed Cu, added a small amount of Ti, and formed a body-centered cubic (BCC) high-entropy alloy, which has very high strength, as high as

DOI: 10.1201/9781003319986-1

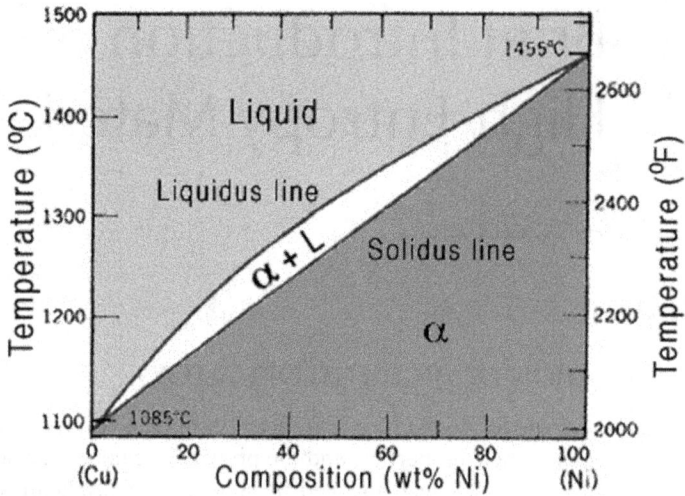

FIGURE 1.1   Metallographic diagram of CuNi alloy.

TABLE 1.1
Sorting and Numbering of Our Research Group's HEAs

Number	Alloy Composition	Phase
GS101	$Al_{0.3}CoCrFeNi$	FCC
GS102	$Fe_{28.5}Co_{47.5}Ni_{19}Al_{1.6}Si_{3.4}$	FCC
GS201	$AlCoCrFeNiTi_{0.2}$	$BCC + B_2$
GS202	$W_{0.2}Ta_{0.2}FeCrV$	BCC
GS203	$Zr_{45}Ti_{31.5}Nb_{13.5}Al_{10}$	BCC
GS301	$AlCo_{0.4}CrFeNi_{2.7}$	$FCC + B_2$

Note: GS is the initial letter of high-entropy Chinese; the first number defines the phase structure of the alloy (for example, (1) FCC, (2) BCC, and (3) duplex or polyphase); the final two digits reflect the alloy development sequence (for example, the alloy first evolved inside the FCC-structure group: GS101).

3000 MPa, and excellent compressive ductility, which can beat a lot of high-strength bulk amorphous alloys. Up to now, lots of high-entropy alloys exhibiting excellent properties have been reported. Typically, they can break the trade-off between high strength and elongation, they can break irradiation resistance, they can be corrosion resistant, and they can have very excellent soft magnetic properties with very high ductility for shaping and deformations (Table 1.1).

1.2 ENTROPY

Entropy is a measure of the disorder in a system. Its concept was first proposed by Closius. He found that the heat usually was driven by the temperature gradient and flow from the high-temperature region to the lower-temperature region, and the diffusion is usually from the high-concentration region to the lower-concentration region, which is driven by the concentration gradient. Then Boltzmann proposed the configuration entropy, which related to the configuration microstates:

$$S = kLnW \qquad [1.1]$$

Here, k is the Boltzmann constant, which is 1.38×10^{-23} J/K, and W is the configuration microstates corresponding to the macrosystem. For the regular solution, the equation can be simplified as follows:

$$S = -R\Sigma XiLnXi \qquad [1.2]$$

Here, R is the gas constant, 8.31 J/Kmol; Xi is the concentration of the ith component.

Based on 1.2, Shannon proposed the Shannon entropy, which is mainly used in information systems.

$$S = -\Sigma XiLnXi \qquad [1.3]$$

Although diffusion is usually driven by the concentration gradient, lots of phenomena are related to the diffusion from the low concentration to the high concentration. This behavior is related to precipitations and spinodal decomposition; thus, the processing is not only controlled by entropy but also enthalpy, which is related to atomic bonding, attractive or repulsive; hence, the following Gibbs free energy:

$$G = H - TS \qquad [1.4]$$

Here in the equation, H is enthalpy, S is entropy, and T is the absolute temperature.

The processing is usually a balance between the effect of entropy and enthalpy. Because we are considering a mixing process, we can mix the particles or the components, which can be gas state, liquid state, or solid state, even plasma state. We consider the entropy and enthalpy changes; then we can define a new parameter, which can evaluate the entropy effect over the enthalpy effects:

$$\Omega = Tm\Delta S/\Delta H \qquad [1.5]$$

Here, Tm is the average melting temperature, ΔS is the entropy of mixing, and ΔH is the enthalpy of mixing.

The component sizes also have a great effect on the phase and structure formation (Figure 1.2 and Figure 1.3). Here, for the elemental component, we can use the atomic radium difference δ, which can be defined as the following equation:

$$\delta = \sqrt{\Sigma xi(1 - Ri/R0)2} \qquad\qquad [1.6]$$

We have summarized the phase formation regions with the parameters of ΔH and δ, and the solid-solution phase formed at a region with a smaller value of absolute ΔH and smaller δ, while the amorphous phase formed at a region with negative ΔH and larger values of δ. If we only consider the solid solutions, the FCC phase usually has a smaller value of δ than that of the BCC phases.

1.3   CORE EFFECTS

The core effects of high-entropy alloys have been summarized as (1) thermodynamically high entropy, (2) kinetically sluggish diffusion, (3) structural distortion severely, and (4) cocktail properties. These four effects may not be applicable in all the high-entropy alloys and materials, but they are usually correct. There are also some other characteristics discovered by recent studies. We can summarize them

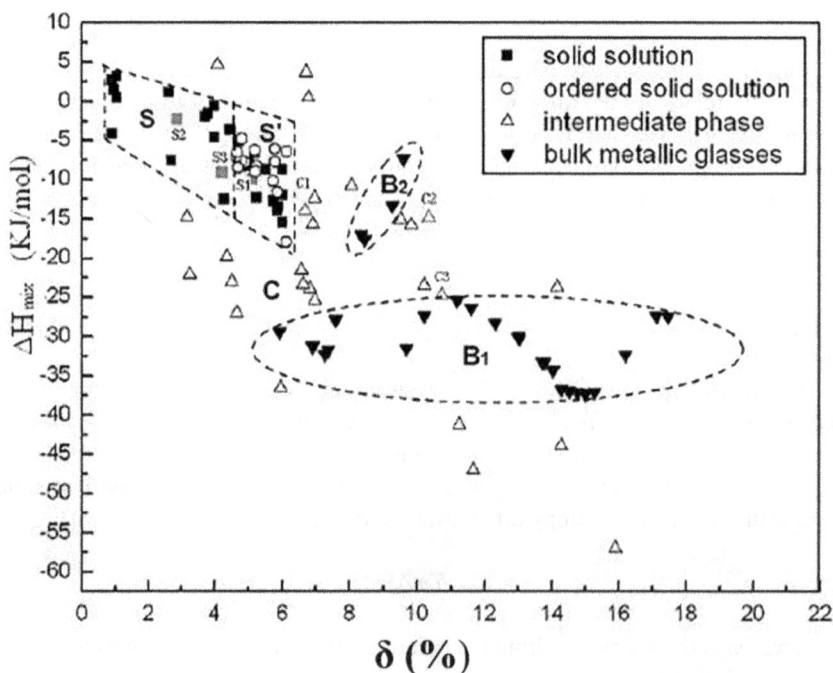

FIGURE 1.2   Relationship between ΔH and smaller δ and the phase area diagram.

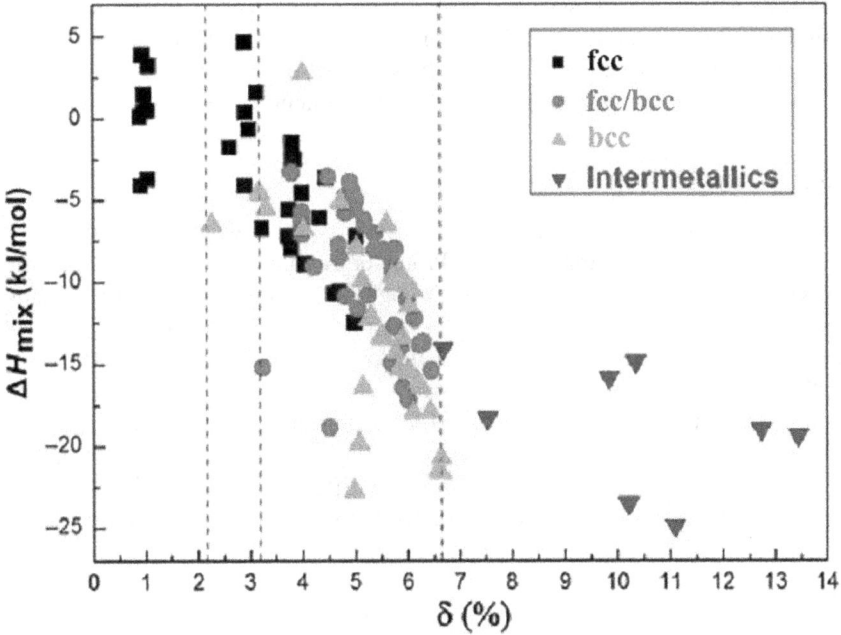

FIGURE 1.3 Relationship between ΔH and smaller δ and the resulting crystal structure.

as follows: (1) ultrahigh strength, (2) excellent irradiation resistance, (3) very low ductile-brittle transition temperature (DBTT), (4) very high strength at elevated temperatures, (5) very high corrosion resistance, and (6) breaking the trade-off between strength and ductility.

1.3.1 HIGH-ENTROPY EFFECT

The high-entropy effect is the most important characteristic of a high-entropy alloy. According to Boltzmann's hypothesis on the relationship between entropy change and system chaos, the molar entropy change (coordination entropy) generated when N equimolar elements are mixed to form solid melt is $\Delta Sconf = R \cdot ln(n)$, so when $n = 2, 3$, and 5, δS is $0.69R$, $1.10R$, and $1.61R$, respectively. Because the traditional alloy is dominated by one element, the mixing entropy of the alloy is generally below $0.693R$, which is far lower than the limit of $1.5R$ of the high-entropy alloy. It can be seen that the greater the principal element number of the alloy, the greater the entropy value of the alloy system and the smaller the free energy of the system. When the mixing entropy of the alloy system is greater than the entropy change of intermetallic compounds, the compatibility of alloying elements in the system increases, and the solid solution with a simple structure of FCC, BCC, or HCP phases tends to be formed. The high mixing entropy makes the alloy tend to form solid solutions rather than intermetallic compounds. This is because intermetallic compounds are ordered phases, and the configuration entropy is approximately zero due to their continuous chemical composition and specific lattice structure. Figure 1.4 shows the relationship between the entropy of mixing $\Delta Sconf$ and the number of components n of the alloy with an equal substance ratio [1]. It can be seen that the $\delta Sconf$ value of the system increases with the increase of the number of components n.

1.3.2 SLOW DIFFUSION EFFECT

Many studies have shown that the self-diffusivity of elements in multi-principal alloys is one order of magnitude lower. Professor Ye Junwei [2] designed three diffusion couples of CrMn, FeCo, and FeNi through pseudo-binary alloy for verification. The Q/T value of Cr, Mn, Fe, Co, and Ni in CrMnFeCoNi high-entropy alloy is the largest; that is, the diffusion coefficient is the lowest. This is the most direct evidence for the slow diffusion effect in high-entropy alloys. Because high-entropy alloys in the interaction between atoms and the lattice distortion affect the effective diffusion rate of atoms, and atoms in the high-entropy alloys were mainly spread through the vacancy mechanism due to the melting point of different atoms of different sizes and bonding strength, strong activity is more likely to spread to open position; if filled,

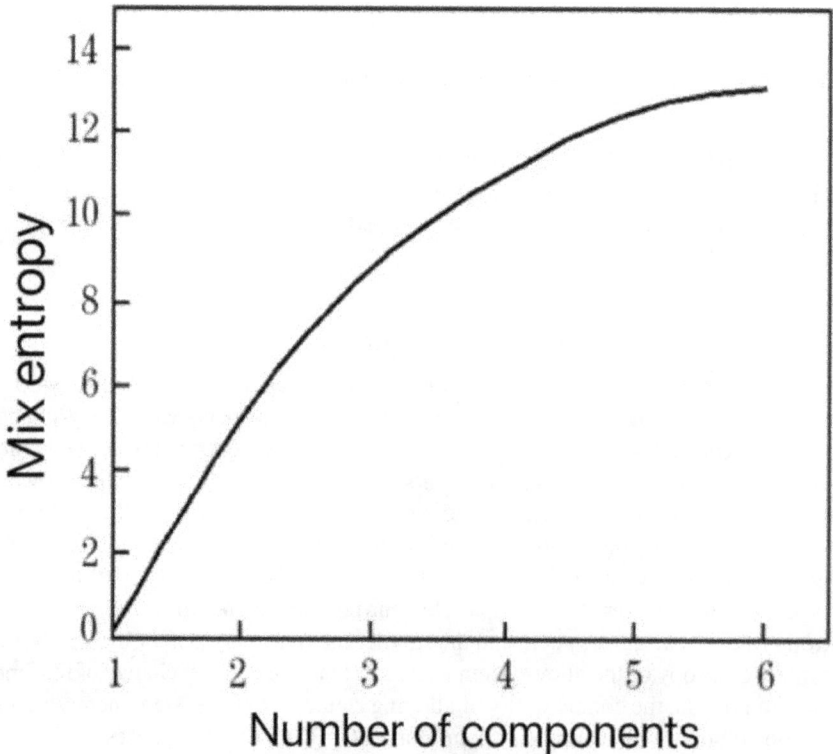

FIGURE 1.4 The relationship between the entropy of mixing $\Delta Sconf$ and the number of components N of an alloy with equal substance ratio.

the vacancy of atomic energy decreases after hardening and continues to spread. If the energy increases, it is difficult to enter the vacancy, so the diffusion rate and transformation rate of the high-entropy alloy solid solution decreases.

This diffusion in high-entropy alloys is like traffic at the intersections in our lives, which is prone to congestion due to the increasing number of vehicles, causing cars to move more slowly (Figure 1.5).

1.3.3 Lattice Distortion Effect

High-entropy alloy is a single-phase solid-solution structure mixed by a variety of elements with an equal atomic ratio. The crystal structure is distorted by the difference in atomic size, bond type, and lattice potential energy, which is the lattice distortion of high-entropy alloy. The lattice distortion effect is often closely related to the structure and mechanical properties of alloys, and it is also the most important core effect in high-entropy alloys.

In 2004, Yeh et al. [3] characterized lattice distortion in high-entropy alloys by X-ray method for the first time and believed that large lattice distortion would increase the degree of X-ray scattering. In 2017, Owen used strong penetration neutron diffraction to characterize six single-phase FCC solid-solution alloys: Ni, $Ni_{80}Cr_{20}$, $Ni_{75}Cr_{25}$, $Ni_{67}Cr_{33}$, $Ni_{37.5}Co_{37.5}Cr_{25}$, and FeCoNiMnCr. Zhang et al. studied the lattice distortion caused by adding a large number of Pd atoms to FeCoNiCr in

FIGURE 1.5   Traffic at intersections will be congested due to the increase in the types of vehicles.

Source: Courtesy of weibo.com/sjgd2013

detail by high-energy X-ray diffraction method and found that, with the increase of Pd atoms, the effect of lattice distortion is gradually significant.

The relationship between εRMS and the phase stability of hundreds of alloy components is summarized in Figure 1.6 [4]. As can be seen in the figure, the formation condition of a single-phase solid solution is εRMS < 5%, and with the increase of residual stress, the polyphase structure appears to release the residual stress, so the formation condition of the polyphase structure is 5% < εRMS < 10%. When εRMS > 10%, the crystal structure will form a disordered amorphous structure due to excessive lattice distortion.

1.3.4 COCKTAIL EFFECT

High-entropy alloys consist of many elements, and each element has different characteristics. The interaction between different elements makes the alloy show a compound effect; that is, a cocktail effect, which was first proposed by Indian scholar Ranganatha. The cocktail effect of high-entropy alloys is more emphasized on the effect of the alloy's principal elements on the atomic scale, which will eventually affect the macroscopic properties of the alloy and even produce additional effects. As shown in Figure 1.7, in an AlCoCrCuFeNi high-entropy alloy, the formation of the BCC phase is continuously promoted with the increase of Al content, and the hardness of the alloy also changes with the change of phase [5].

High-entropy alloys formed by alloying a variety of elements with specific atomic ratios generally have high hardness, high strength, high corrosion resistance, high

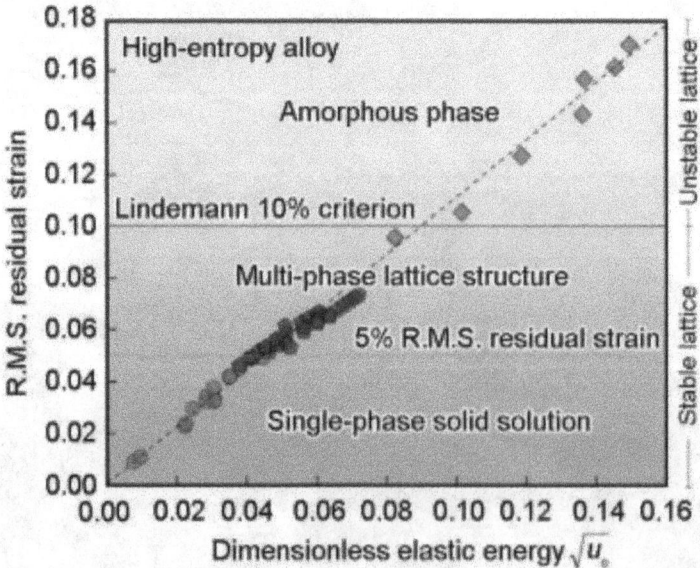

FIGURE 1.6 The correlation between root mean square (RMS) residual strain εRMS and the phase of different alloys.

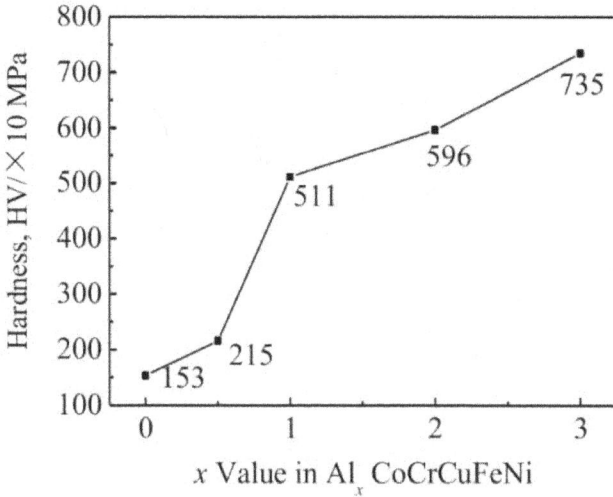

FIGURE 1.7 Hardness of Al$_x$CoCrCuFeNi at different Al content.

temperature stability, and other properties. However, since more research is still in the stage for the multiple-phase diagram of alloy composition design and now for the primary design of high-entropy alloys, hence, just now through the way of cocktail, changing the kind and content of the alloy elements to achieve the requirement of alloy microstructure and performance, the design of high-entropy alloys has not yet formed a scientific theory guidance. For example, if the tensile strength of the alloy is high but the hardness is not special, we will naturally choose the element with FCC structure at the beginning of the alloy design. If the material to be designed will be used in the aerospace field, we will consider lightweight elements as alternatives. If the alloy is required to be used in a high-temperature-resistant environment, refractory elements will be the first choice. Therefore, at the beginning of alloy design, various factors should be considered comprehensively to select the appropriate combination of elements and the corresponding preparation process.

1.4 BREAKING THE TRADE-OFF BETWEEN STRENGTH AND DUCTILITY

Higher strength and better plasticity are the eternal pursuits of structural materials. However, strength and plasticity seem to be a pair of contradictions in alloys. Alloys with higher strength often lack plasticity and vice versa. Recent studies have shown that high-entropy alloys have the advantage of breaking through the strength-plasticity rule of traditional alloys due to their novel and unique composition design concept. The core idea of high-entropy alloys is to stabilize the chemically disordered solid-solution phase by increasing the configurational entropy of alloys and inhibit the formation of ordered intermetallic compounds competing with it. It is expected that the single-phase solid-solution alloy has high strength and good plasticity. Figure 1.8 shows the comparison between the strength and ductility of traditional alloys and high-entropy alloys. It can be clearly seen that high-entropy alloys have good ductility while maintaining high strength compared with traditional alloys, which has the potential to break the balance between the strength and ductility of traditional alloys. Overcoming the strength and ductility balance of high-entropy alloys is more dependent on the optimization of processing conditions and the adjustment of phase composition. The main mechanisms are the high-concentration solid-solution structure, the sensitivity of grain size induced by serious lattice distortion, and the regulation of the stability of the component phase, such as the introduction of nano precipitation and TRIP/TWIP.

Li et al. designed a new high-entropy alloy, $CoCrFeNi_2Al_{0.3}Ti_{0.25}$ (Figure 1.9), to form nano-layered sediments and ultrafine grains to enhance the alloy strength, giving it high strength and excellent ductility. The cold-rolled samples were annealed at 600°C(CR-600), 700°C(CR-700), and 800°C(CR-800) for 1h, and then cooled by water. The mechanical properties test diagram showed that CR-700 had excellent strength and ductility [6].

Xu et al. [7] designed a copper base high-entropy alloy based on the CoCrFeNi system high-entropy alloy by adding Cu. They systematically studied the microstructure changes of biphasic (Cu-rich and CoCrFeNi-rich) face-centered cubic $CoCrFeNiCu_4$ alloys during various thermal cycles and thermal mechanical processes. During the heating process, the copper-rich particles preferentially precipitate, resulting in a more obvious composition-difference between the two components of the face-centered cubic phase and an increase in the relative volume fraction of the copper-rich phase.

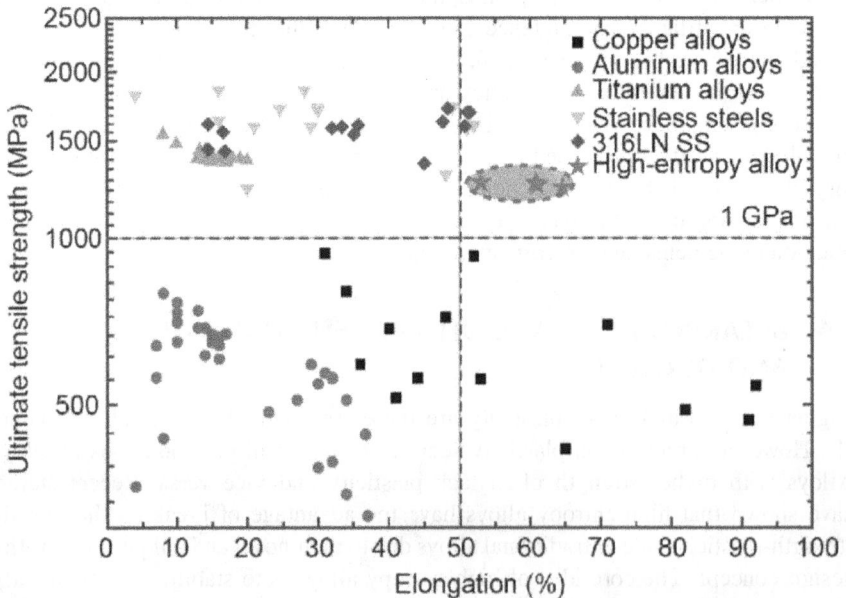

FIGURE 1.8   The relationship between strength and elongation.

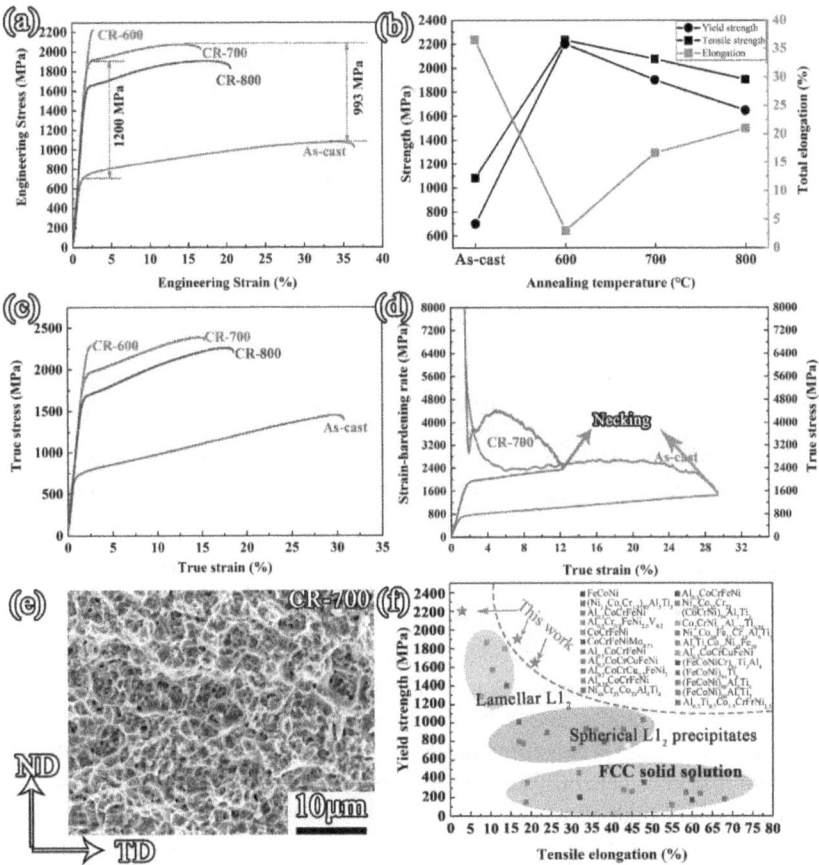

FIGURE 1.9 Mechanical properties of $CoCrFeNi_2Al_{0.3}Ti_{0.25}$ HEAs. (a) Engineering stress-strain curves of the HEAs at room temperature, CR-700 HEA exhibits ultrahigh strength and excellent ductility; (b) variation of yield strength, ultimate tensile strength, and total elongation at different annealing temperatures; (c) true stress-strain curves of HEAs; (d) work-hardening rate curves of as-cast and CR-700 HEA; (e) fracture surface of CR-700 HEA, showing plenty of fine dimples, indicating a characteristic mode of a ductile fracture; and (f) excellent yield strength and total elongation of our present HEAs (CR-600, CR-700, and CR-800) compared with other HEAs, respectively.

The alloy shows continuous melting and discontinuous solidification of rich Cu and CoCrFeNi phases. The alloy shows continuous melting and discontinuous solidification of rich Cu and CoCrFeNi phases. Figure 1.10 shows the mechanical properties of CoCrFeNi-based copper base high-entropy alloy. After cold-rolling to 90%, the alloy exhibits a recrystallization temperature higher than 800°C, and the yield strength reaches 900 MPa. The strength increases further after annealing at 300°C. With the increase of annealing temperature, the strength decreases and the plasticity toughness

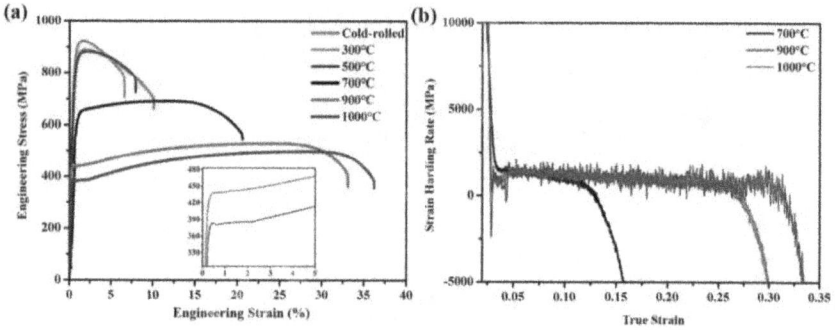

FIGURE 1.10 Mechanical properties of CuCoCrFeNi copper base high-entropy alloy. (a) Engineering stress ~ Engineering strain curves; (b) strain hardening rate ~ true strain curves.

increases gradually. The high-entropy alloys represented by CoCrFeNi have the advantage of breaking through the law of strength-plasticity loss of traditional alloys.

1.5 BREAKING THE PERFORMANCE LIMITS OF TRADITIONAL MATERIALS

Materials with low temperature and high plasticity are needed in many fields, such as space exploration, low-temperature storage, and nuclear reactor. With the progress of nuclear reactors and space technology, the demand for high-performance, low-temperature materials is becoming more and more urgent. However, the plasticity of traditional materials generally decreases with a decrease in temperature. The special design concept of high-entropy alloy gives it the potential to break the performance limit of traditional materials. This also lets people see hope in the field of low temperature and high plasticity.

George's research group [8] systematically studied the low-temperature and high-temperature tensile properties of face-centered cubic high-entropy alloys and found that the CoCrFeNiMn alloy with face-centered cubic structure was stronger and tougher at a lower temperature than room temperature, at 77 K. Zhang Yong's research group [9] studied the mechanical properties of CoCrFeNi high-entropy alloy with face-centered cubic structure at ultralow temperature. It was found that the tensile strength of the alloy reached 1260 MPa at 4.2 K (Figure 1.11), and the elongation reached 62%, showing good low-temperature mechanical properties and great breakthroughs in low-temperature plasticity. By studying the cause of high plasticity at low temperatures, the extremely low lamination energy of high-entropy alloys easily induces deformation twins. The existence of deformation twins can make the alloy maintain high strength and high plasticity. The phase transition of FCC-HCP in the alloy is the cause of plasticity reduction at 77 K temperature.

1.6 BREAKING THE TRADE-OFF BETWEEN THE PHYSICAL PROPERTIES AND MECHANICAL PROPERTIES

The ability of high-entropy alloys to design unique combinations of mechanical and functional properties in an infinite compositional space is encouraging. In general,

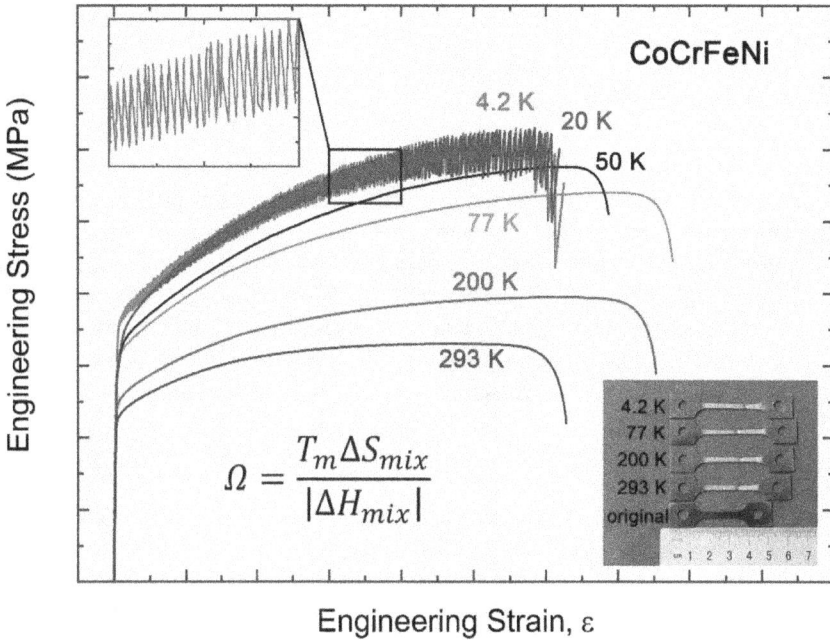

FIGURE 1.11 Tensile stress-strain curves of CoCrFeNi high-entropy alloy at different temperatures.

functional properties are obtained from traditional materials, which, however, do not usually provide good service under extreme conditions and do not simultaneously retain excellent mechanical properties. Poor plasticity at low temperatures, high strength, and poor electrical conductivity are all problems faced by traditional materials. The appearance of high-entropy alloys gradually breaks the balance between physical properties and mechanical properties, and more high-entropy alloys with excellent properties are developed.

The excellent thermal stability of the high-entropy alloy is attributed to its multi-component design. The interaction between different atoms limits the diffusion rate of the high-entropy alloy. BCC high-entropy alloys generally have higher thermal stability than FCC high-entropy alloys. For example, the NbMoTaW high-entropy thin film has better structural and mechanical stability than the traditional W alloy. It is found that the yield strength of WTaFeCrV alloy is higher than that of other alloys at 800°C. The slow diffusion effect and lattice distortion effect of high-entropy alloys can slow down the grain coarsening at a low temperature so that the high-entropy alloy has good thermal stability. According to this property, Zhang et al. [10] prepared FeCoNiCu NC-HEA (80 nm) with excellent thermal stability, which maintained high yield strength after annealing at 900°C.

Huang et al. [11] used first principles to calculate the bedding fault energy of CoCrFeNiMn high-entropy alloy. They divided the bedding fault energy into three parts, including chemical, magnetic, and strain parts, and calculated each part by using structural energy, local magnetic moment, and elastic modulus, which theoretically proved that a high-entropy alloy has lower fault energy. The low lamination fault energy easily induces twins, and the high-density twin boundary can make the alloy have high conductivity while maintaining high strength. It is proven theoretically that the high-entropy alloy has the potential for high strength and high conductivity.

High-entropy alloys will be the focus of research in the future because of their unique trade-off between mechanical properties and physical properties, such as conductivity, strength, and thermal stability. In addition, the future development of high-entropy alloy not only focuses on being performance-driven but also needs to develop new high-entropy materials suitable for deformation, casting, powder making, powder smelting, and other special processes from the perspective of being process driven, which will be of great significance to the further development of high-entropy alloys and the expansion of its application field research.

REFERENCES

[1] Li W H, Ai T T. More main yuan high entropy alloys into research progress [J]. *J. Powder Metall. Ind.* 2016, 26 (1): 64–67.
[2] Yeh J W, Lin S J, Chin T S, et al. Formation of simple crystal structures in Cu-Co-Ni-Cr-Al-Fe-Ti-V alloys with multi principal metallic elements [J]. *Metall. Mater. Trans.* 2004, 35A: 2533.
[3] Yeh J W, Chang S Y, Hong Y D, et al. Anomalous decrease in Xray diffraction intensities of Cu-Ni-Al-Co-Cr-Fe-Si alloy systems with multi-principal elements [J]. *Mater. Chem. Phys.* 2007, 103: 41.
[4] Ye Y F, Liu C T, Yang Y. A geometric model for intrinsic residual strain and phase stability in high entropy alloys [J]. *Acta Mater.* 2015, 94: 152–161.
[5] Liu Y, Chen M, Li Y X, Chen X. Microstructure and mechanical performance of AlxCoCrCuFeNi multi-principal high entropy alloy [J]. *Rare Metal Mat. Eng.* 2009, 38(09): 1602–1607.
[6] Li W, Chou T H, Yang T, Chuang W S, Chen F R. Design of ultra-strong but ductile medium-entropy alloy with controlled precipitations and heterogeneous grain structures. *Appl. Mater. Today.* 2021, 23: 101037.
[7] Xu Z, Li Z, Tong Y, Zhang W, Wu Z. Microstructural and mechanical behavior of a CoCrFeNiCu4 non-equiatomic high entropy alloy [J]. *J. Mater. Sci. Technol.* 2021, 60: 35–43.
[8] Gali A, George E P. Tensile properties of high- and medium-entropy alloys [J]. *Intermetallics.* 2013, 39: 74–78.
[9] Liu J, Guo X, Lin Q, He Z, An X, Li L, Liaw P K, Liao X, Yu L, Lin J, et al. Excellent ductility and serration feature of metastable CoCrFeNi high-entropy alloy at extremely low temperatures. *Sci. China Mater.* 2018, 62: 853–863.
[10] Zhang Y T, Liu M W, Sun J D, Li G D, Zheng R X, Xiao W L, et al. Excellent thermal stability and mechanical properties of bulk nanostructured FeCoNiCu high entropy alloy. *Mater. Sci. Eng.: A* 2022, 835: 142670.
[11] Huang S, Li W, Lu S, Tian F, Shen J, Holmström E, Vitos L. Temperature dependent stacking fault energy of FeCrCoNiMn high entropy alloy [J]. *Scr. Mater.* 2015, 108: 44–47.

2 FCC-Structured High-Entropy Materials

Yong Zhang and Changwei Li

2.1 INTRODUCTION

FCC-structured high-entropy alloys usually exhibit low strength at the casting state but ductile. They can be deformed by cold-rolling and hot-forging; the fracture toughness and impact toughness of the FCC high entropy are usually extraordinarily high; precipitation and grain refinements; alloys are designed by using machine learning and high-throughput screening experiments; typical alloys are, for example, GS101 ($Al_{0.3}CoCrFeNi$) and GS102 ($Co_{47.5}Fe_{28.5}Ni_{19}Si_{3.4}Al_{1.6}$).

The face-centered cubic structure of high-entropy alloys is similar to that of traditional alloys, except that different atoms tend to occupy the lattice randomly, which causes more severe lattice distortion. When the atoms are randomly arranged in a high-entropy alloy, the disordered FCC structure (A1 structure) is formed. When the interatomic interaction in the alloy is very strong and ordered, as in the L12 structure, most of the face-center positions are occupied by a particular metal atom, and the lattice vertex positions are occupied by other atoms. Compared with traditional L12 structure, the order degree of L12 structure in high-entropy alloy is slightly decreased.

2.2 GS101

The $Al_{0.3}CoCrFeNi$ high-entropy alloy, with a simple crystal structure, good machinability, and ultrahigh plasticity, has been a wide concern and has been studied by researchers in recent years.

Recently, a number of researchers have conducted research on GS101 rolling, heat treatment, and other processes to explore whether the performance of GS101 can be optimized through these processes. Tang Qunhua et al. studied the microstructure and mechanical properties of the FCC $Al_{0.3}CoCrFeNi$ high-entropy alloy after 90% reduction rolling and annealing [1]. The results show that the alloy recrystallizes after rolling and annealing (600°C~1000°C), and the ordered BCC phase enriched with Al and Ni atoms is preferentially formed at the grain boundary of the recrystallized FCC phase, and its volume fraction increases first and then decreases with the increase of annealing temperature. The alloy was significantly strengthened by rolling, and then annealing at 600°C could further strengthen the alloy without sacrificing the uniform plasticity. Increasing the annealing temperature would lead to a decrease in strength and an increase in the plasticity of the alloy. After annealing at 800°C, the alloy exhibits ideal strength-plasticity matching, with uniform elongation of 34.1%

DOI: 10.1201/9781003319986-2

and tensile strength of 935 MPa, which is about three times that of as-cast alloy (303 MPa). This is mainly attributed to the refinement of the recrystallization structure and precipitation strengthening of the ordered BCC phase.

Sun Yongzhe et al. studied the effect of annealing temperature on the microstructure and mechanical properties of the $Al_{0.3}CoCrFeNi$ high-entropy alloy with 95% cold-rolling deformation [2]. The results show that the alloy remains at FCC single phase after 95% cold-rolling deformation. After cold-rolling, the hardness of the alloy is obviously improved, the plasticity is greatly reduced, and the strength is increased by 4~5 times. After annealing at 600°C, the alloy recrystallized. With the increase of annealing temperature, the grain size increases gradually, the strength of the alloy decreases, the plasticity increases, and the fracture morphology changes from cleavage to dimple. At 600°C~800°C, a small amount of second-phase (BCC-phase) precipitates in the alloy. The higher the temperature is, the more obvious the second-phase precipitates.

Si songhua et al. studied the effects of different cooling methods (air cooling and furnace cooling) on the microstructure and mechanical properties of the $Al_{0.3}CoCrFeNi$ high-entropy alloy after cold-rolling at 75% reduction and heat treatment at 1073 K for 1 h [3]. The results show that the as-cast and cold-rolled $Al_{0.3}CoCrFeNi$ alloys have FCC single-phase structure, and the stove-cooled and air-cooled $Al_{0.3}CoCrFeNi$ alloys have FCC+BCC dual-phase structure after heat treatment. After cold-rolling, the strength of as-cast alloy increases significantly, but the ductility decreases greatly. Due to fine-grain strengthening, twin strengthening, and precipitated phase strengthening, the stove-cooled alloy has good comprehensive mechanical properties after heat treatment. Its tensile strength is 1289 MPa, which is about twice that of the as-cast specimen (719 MPa), and its maximum elongation is 28.7%. Due to the increase of precipitated phase and twin size, the tensile strength of the furnace-cooled alloy is increased without loss of plasticity compared with air-cooled alloy.

Li Dongyue et al. prepared an alloy ingot with nominal composition $Al_{0.3}CoCrFeNi$ (atomic percentage) by melting a pure metal mixture (purity > 99% wt%, weight percentage) in a high-purity argon atmosphere with the vacuum suspension method [4]. To ensure chemical uniformity, the ingots were remelted at least three times and then hot-forged and hot-spun into 6 mm bars at 1050°C and 1000°C, respectively. An HEA fiber with a diameter of 1 to 3.15 mm was prepared by thermal method at 900°C. The microstructure was characterized by scanning electron microscopy (SEM) and transmission electron microscopy (TEM). The composition changes within and between fibers was determined by transmission electron microscopy energy dispersive X-ray spectroscopy and atomic probe tomography (APT). These analyses reveal a uniform face-centered cubic (FCC) structure in the as-cast material, and subsequent treatments, such as forging and wiredrawing, produce nanosized B2 particles in the FCC matrix. Electron backscatter diffraction (EBSD) was used to determine the evolution of the fiber texture and grain boundary characteristics after processing. The tensile strength and ductility of the fibers were measured at 298 K (1207 MPa/7.8%) and 77 K (1600 MPa/17.5%). Detailed TEM analysis shows that the improvement in mechanical properties (strength and ductility) at 77 K is due to the transformation of the deformation mechanism from dislocation plane slip to nano-twins. This property may be beneficial for cryogenic applications.

Zhang Yong's research group conducted forging treatment on as-cast $Al_{0.3}CoCrFeNi$ and then conducted a series of performance tests (Figure 2.1(a), Figure 2.1(b), Figure 2.2, and Figure 2.3). The effects of forging deformation on the microstructure and mechanical properties of the $Al_{0.3}CoCrFeNi$ high-entropy alloy were investigated by X-ray diffractometry, optical microscopy, scanning electron microscopy, hardness test, and tensile test. Tensile tests were carried out at room

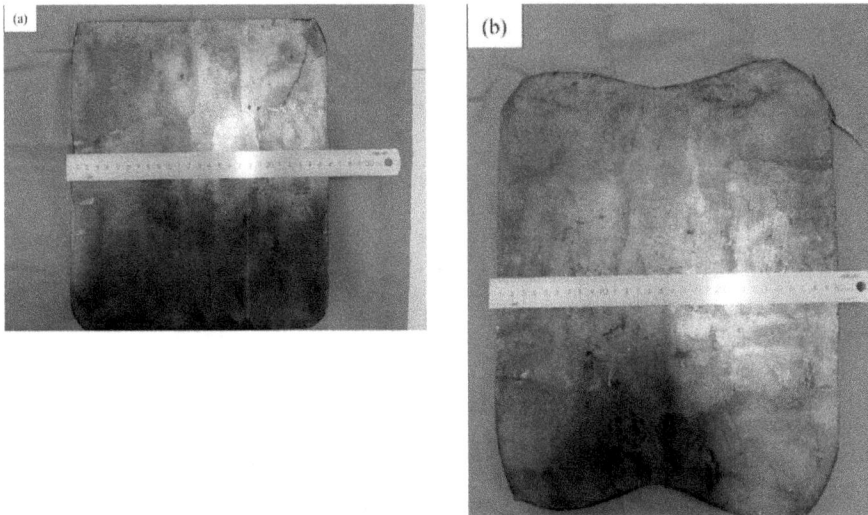

FIGURE 2.1   Forged $Al_{0.3}CoCrFeNi$ plate. (a) plate-1, (b) plate-2

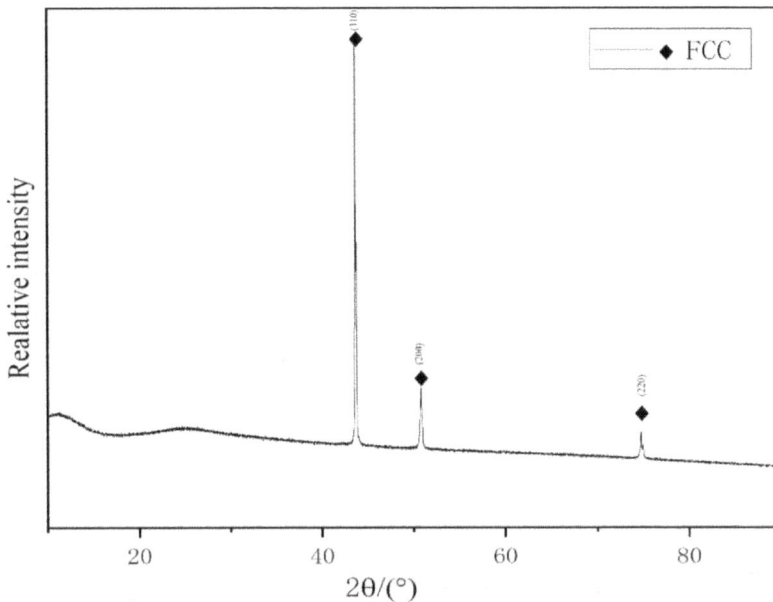

FIGURE 2.2   Forging state $Al_{0.3}CoCrFeNi$ XRD figure.

temperature, low temperature, and high temperature: at 0°C, −10°C, −50°C, −100°C, and liquid nitrogen temperature for low-temperature tensile tests; at 100°C, 300°C, 500°C, 600°C, and 700°C for high-temperature tensile tests. The results show that with the increase of temperature, the tensile strength of the material decreases, the

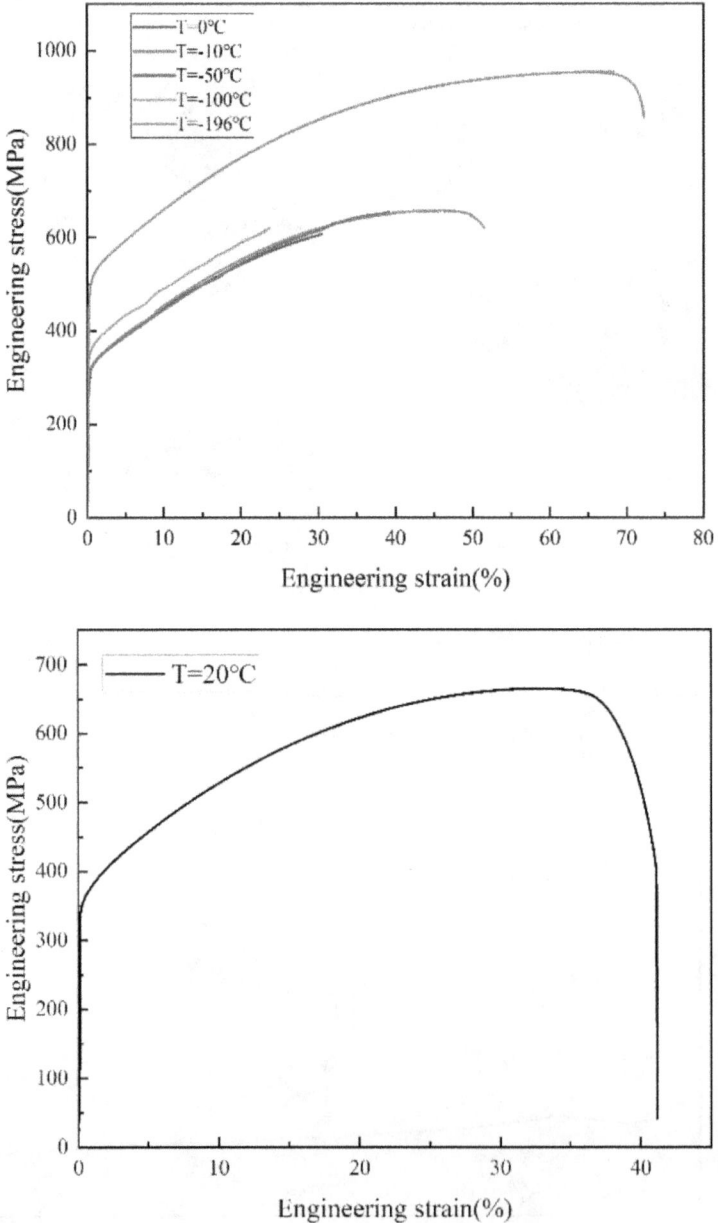

FIGURE 2.3 Low-temperature, room-temperature, and high-temperature tensile test engineering stress-strain curve.

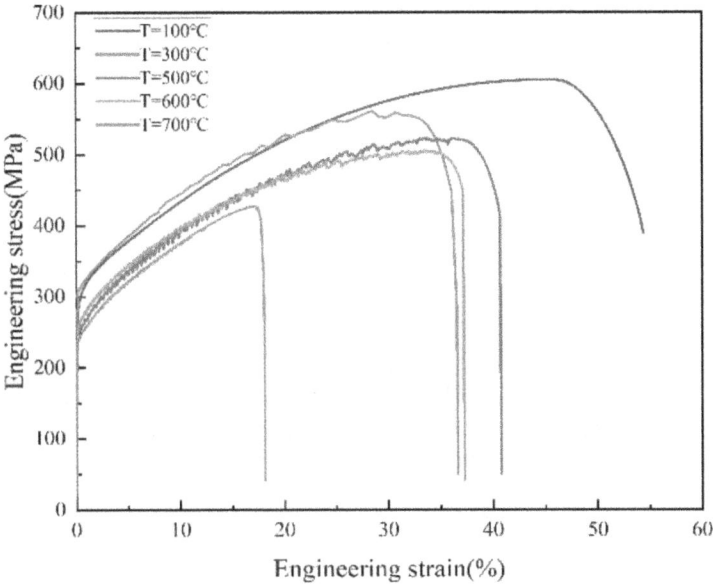

FIGURE 2.3   (Continued)

elongation rate does not change significantly at 20°C–600°C, the elongation rate is the lowest at 700°C, the comprehensive mechanical property is the worst, the tensile strength is 427 MPa, the elongation rate is 27%. With the increase in temperature, the tensile strength of the material increases, and the elongation rate does not change significantly at −100°C–0°C. The elongation rate is the highest at liquid nitrogen temperature, and the comprehensive mechanical properties are the best, in which the tensile strength is 954.2 MPa and the elongation rate is 72.2%.

2.3 GS102

The development of soft magnetic alloys with high cost performance and excellent mechanical properties is of great significance to the energy saving industry. Min zhang et al. investigated the magnetic and mechanical properties of a series of $(Fe_{0.3}Co_{0.5}Ni_{0.2})_{100-x}(Al_{1/3}Si_{2/3})_x$ high-entropy alloys (HEAs) (x = 0, 5, 10, 15, 25) at room temperature [5]. The $Fe_{0.3}Co_{0.5}Ni_{0.2}$-based alloy was chosen because of its lowest saturation magnetostriction coefficient. The results show that the $(Fe_{0.3}Co_{0.5}Ni_{0.2})_{95}(Al_{1/3}Si_{2/3})_5$ alloy retains a simple face-centered cubic (FCC) solid-solution structure after annealing in as-cast, cold-rolled, and 1000°C. After annealing, the tensile yield strength of the alloy is 235 MPa, the ultimate tensile strength is 572 MPa, the elongation is 38%, the saturation magnetization (Ms) is 1.49 T, and the coercivity is 96 A/m. Compared with silicon steel and amorphous soft magnetic materials, the alloy not only has the best combination of soft magnetic and mechanical properties but also has the advantages of easy processing and high thermal stability. Therefore, the $(Fe_{0.3}Co_{0.5}Ni_{0.2})_{95}(Al_{1/3}Si_{2/3})_5$ alloy has good potential as a next-generation soft magnet for a wide range of industrial applications.

Zuo studied the CoFeNi(AlSi)$_x$ series alloy [6]. The research results showed that the saturation magnetization of the alloy decreased linearly with the increase of non-ferromagnetic elements Al and Si, while the yield strength and hardness of the alloy increased continuously. However, excessive Al and Si($x > 0.5$) made the plasticity of the alloy deteriorate. In Ting, the CoFeNi(AlSi)$_{0.2}$ alloy sintered by vacuum arc has good comprehensive properties, such as high ductile saturation magnetization (1.15 T), high resistivity (69.5 μωcm), but high coercivity (1400 A/m). The as-cast alloy inevitably has some defects in microstructure and structure, such as dendrite segregation and holes, and there will be internal stresses, which will affect the properties of the alloy. Therefore, copper die-casting, heat treatment, and the Bridgman directional solidification technology were adopted to improve the magnetic properties of the alloy, and the coercivity of the alloy was significantly reduced. Compared with high GUI steel, the CoFeNi(AlSi)$_{0.2}$ alloy has good soft magnetic properties and easy processing, showing a good application prospect (Figure 2.4, Figure 2.5, Figure 2.6, Figure 2.7, and Figure 2.8).

2.4   OTHER FACE-CENTERED CUBIC HIGH-ENTROPY ALLOYS

Except for GS101 and GS102, most CoCrFeNi series high-entropy alloys have face-centered cubic structures. Many researchers have studied these alloys in depth.

Tao-Tsung Shun et al. studied the effect of Ti addition on the microstructure and mechanical properties of multi-principal component of the CoCrFeNiTi$_x$ (x = 0, 0.3, and 0.5) alloy (Figure 2.9 and Figure 2.10) [7]. The quaternary CoCrFeNi alloy exhibits a simple face-centered cubic solid-solution crystal structure, while the

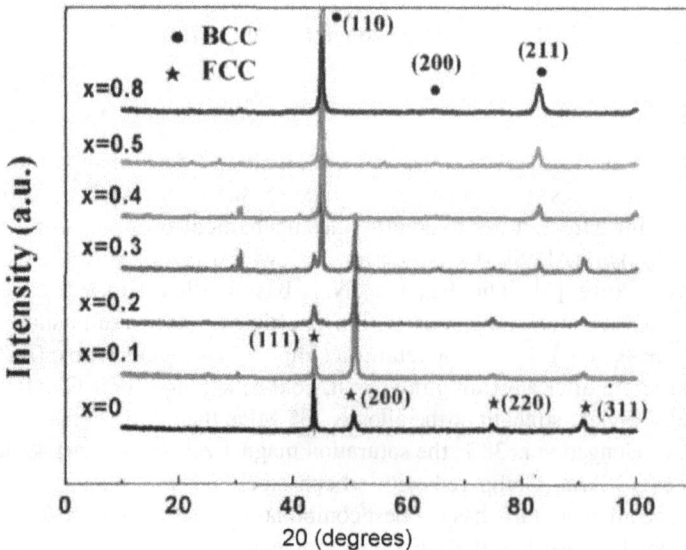

FIGURE 2.4   XRD patterns of high-entropy CoFeNi(AlSi)$_x$ alloys prepared by arc melting.

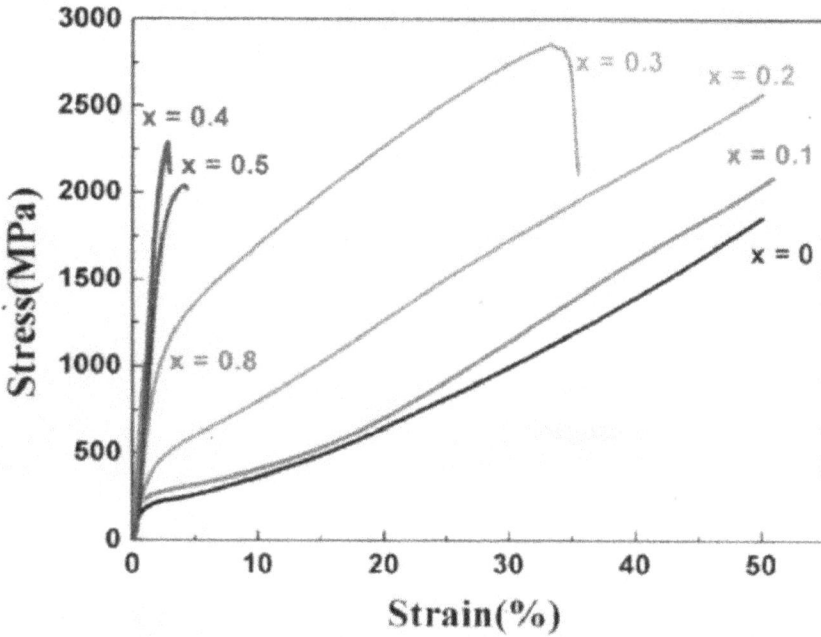

FIGURE 2.5 Engineering stress-strain curve of CoFeNi(AlSi)$_x$ alloy under room-temperature compression.

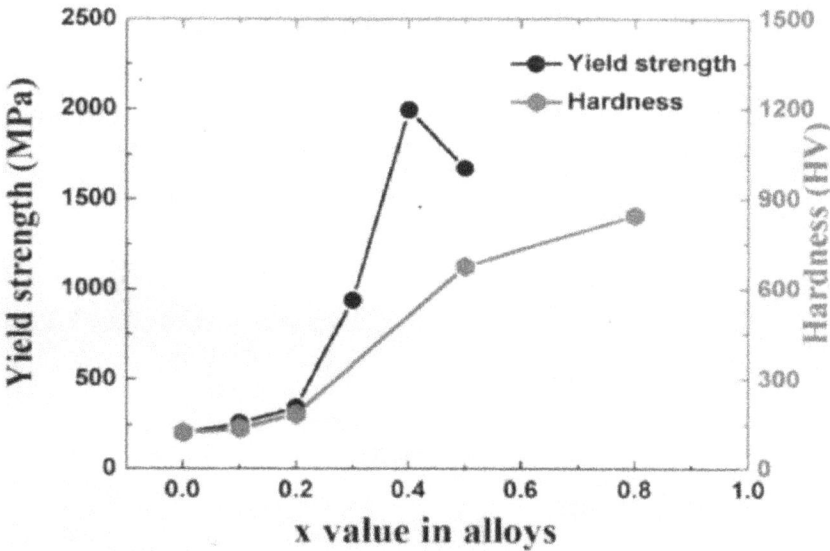

FIGURE 2.6 CoFeNi(AlSi)$_x$ series high-entropy alloys prepared by the arc melting process need variable recognition of Vickers microhardness and yield strength components.

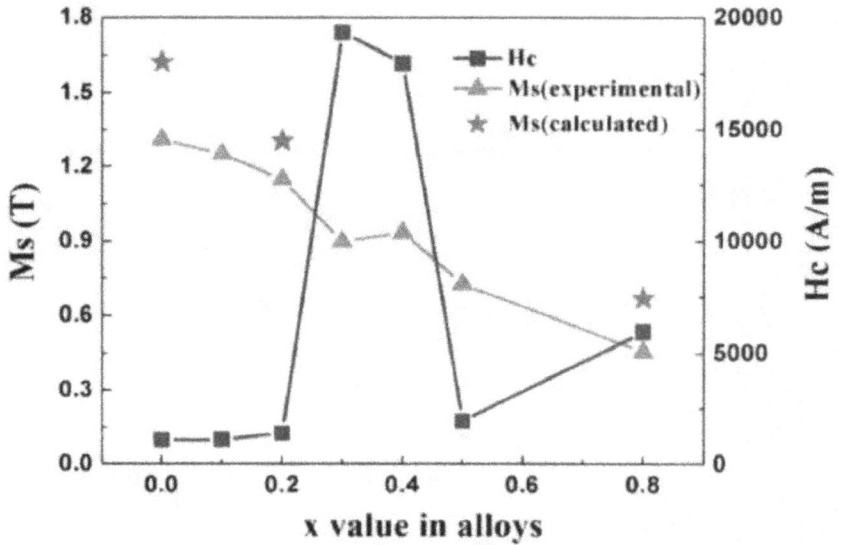

FIGURE 2.7 Coercivity and saturation magnetization of CoFeNi(AlSi)$_x$ series high-entropy alloys prepared by arc melting as a function of gold composition.

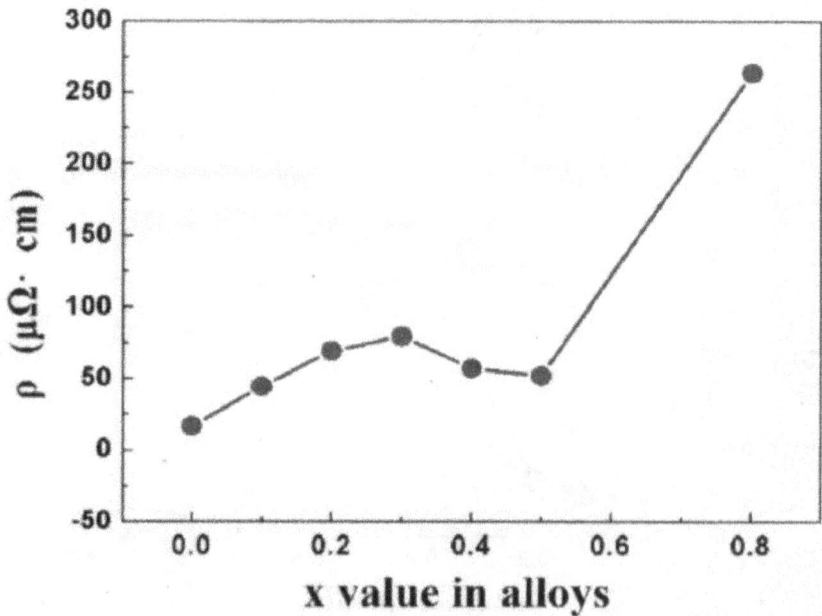

FIGURE 2.8 Composition variation of resistivity of CoFeNi(AlSi)$_x$ series high-entropy alloys prepared by arc melting.

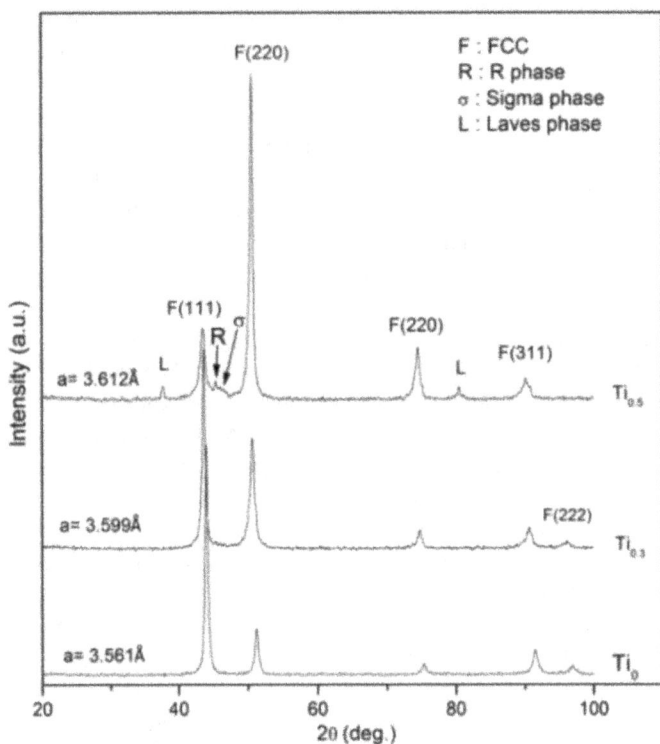

FIGURE 2.9 XRD patterns for the CoCrFeNiTi$_x$ alloys.

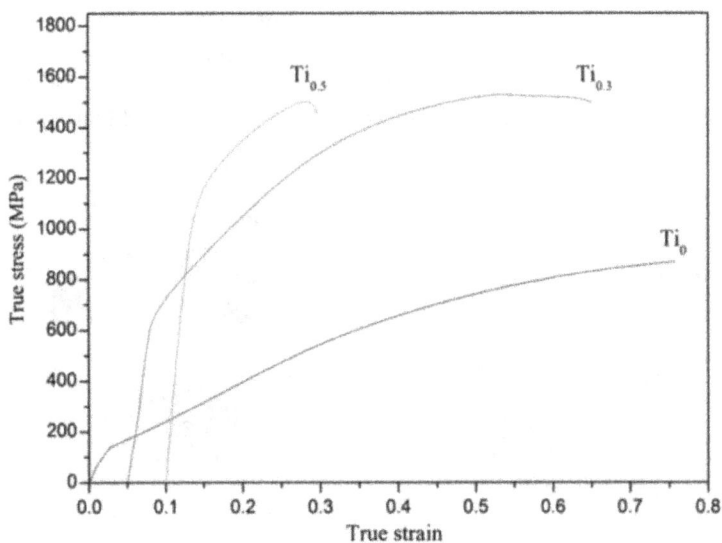

FIGURE 2.10 Compressive stress-strain curves for the CoCrFeNiTi$_x$ alloys.

face-centered cubic matrix of CoCrFeNiTi$_{0.3}$ alloy exhibits a mixed plate-like structure with rich (Ni, Ti) R phase and rich (Cr, Fe) S phase. Face-centered cubic matrix
(Ti, Co) rich Laves phase and Rþ S mixed phase were found in the CoCrFeNiTi$_{0.5}$
alloy. The compressive strength of the alloy increased by about 75% after Ti was
added. The alloy with high Ti content has higher yield stress and lower plasticity.
Solution strengthening and secondary phase hardening of the face-centered cubic
matrix are two main factors of alloy strengthening. The compressive strength and
fracture strain of the CoCrFeNiTi$_{0.3}$ alloy are 1529 MPa and 60%, respectively. This
indicates that the material has the potential to develop into ductile high-strength
alloys.

Anmin Li et al. studied the effects of different forging temperatures on the microstructure, mechanical properties, and wear resistance of FeMnCrCoNi high-entropy
alloys by using magnetic levitation vacuum melting and low-temperature forging
techniques [8]. The results show that the as-cast FeMnCrCoNi high-entropy alloy is
mainly composed of a dendrite phase and black speckle phase, and has an FCC structure. The microhardness and tensile strength of as-cast, room-temperature forged,
and 200°C forged alloys are 158.7, 313.6, and 358.1HV, respectively, and 539.45,
695.45, and 830.84 MPa, respectively. The hardness of as-cast alloy at room temperature and 200°C is 97.6% and 125.6% higher than that of as-cast alloy, respectively.
The tensile strength increased by 28.9% and 54.0%, respectively. The elongation at
room temperature decreased significantly from 47.2% to 15.2% and 5.4% at 200°C,
respectively. The wear resistance of a wrought alloy is better than that of a cast alloy.
The wear resistance order of the alloy is as follows: forging at room temperature is
the best, forging at 200°C is the next, and the as-cast state is the worst.

In the course of the experiment, JiangLi et al. prepared a large-size CoCrFeNiTi$_{0.5}$
alloy ingot in an intermediate-frequency induction-melting furnace (Figure 2.11 and
Figure 2.12) [9]. There is a slight volume effect, which is mainly manifested by the
growth of grains and the increase in the hardness of the alloy (Table 2.1). In order to
investigate the effect of annealing temperature on the microstructure and properties
of the CoCrFeNiTi$_{0.5}$ alloy, the annealing temperatures of 600°C, 700°C, 800°C,
and 1000°C for 6 h were, respectively, carried out. It was found that the annealing

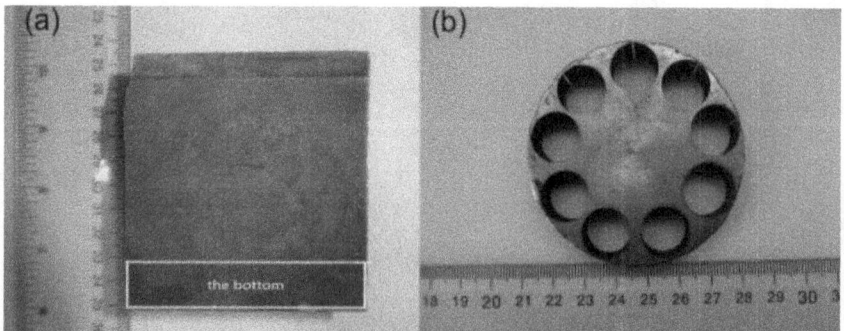

FIGURE 2.11 The position schematic diagram of the samples from the big ingot: (a) the
longitudinal section of the big ingot and (b) the cross-section of the bottom.

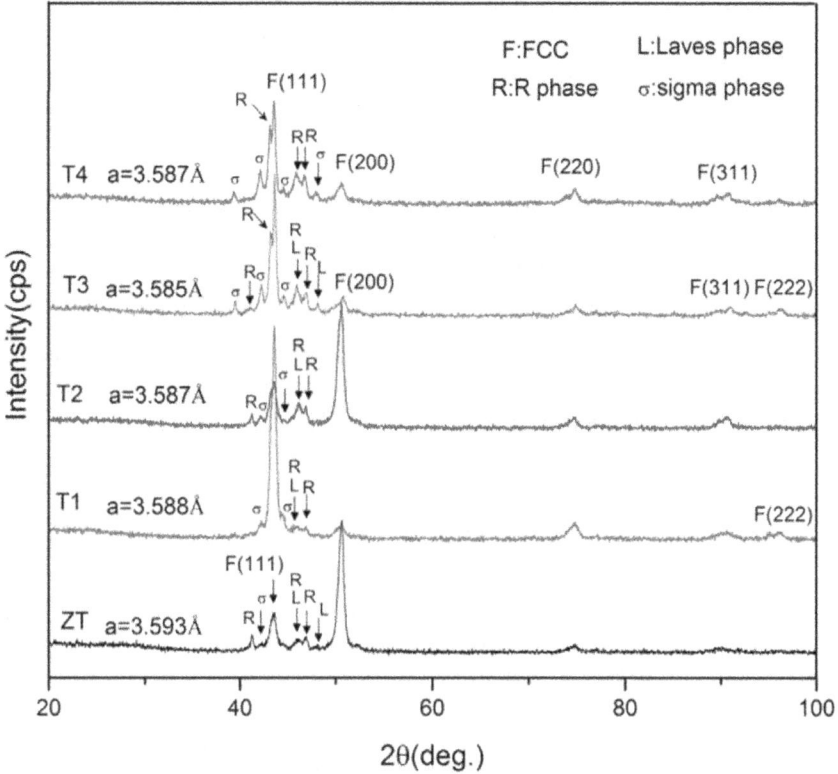

FIGURE 2.12 XRD patterns of the as-cast and annealed CoCrFeNiTi$_{0.5}$ alloy.

TABLE 2.1
Alloy Microhardness, DR Hardness, ID Hardness, and Macrohardness of the CoCrFeNiTi$_{0.5}$ Alloy in Various Heat Treatments

	ZT	T1	T2	T3	T4
Alloy hardness (HV)	616.80	532.58	561.34	528.52	511.52
DR hardness (HV)	581.42	495.70	516.30	505.12	462.20
ID hardness (HV)	751.28	591.42	589.68	631.78	563.92
Macro hardness (HRC)	52.00	46.67	54.25	49.25	48.50

temperature below 1000°C had almost no effect on the microstructure and element distribution of the alloy. The crystal structure of the CoCrFeNiTi$_{0.5}$ alloy is composed of a main face-centered cubic (FCC) solid-solution matrix and a small number of intermetallic phases in the form of interdendritic crystals. The content of Co, Cr, Fe, and Ni in dendrites is similar, while the content of Ti is low. When annealed below 1000°C, the interdendritic crystals are in the rich (Ni, Ti), rich (Fe, Cr), and rich (Co, Ti) phases. After annealing at 1000°C, the (Co, Ti) rich phase disappears, and the (Ni, Ti) rich phase and (Fe, Cr) rich phase grow. The microhardness and macrohardness of the as-cast CoCrFeNiTi$_{0.5}$ alloy are 616.80 HV and 52 HRC, respectively. After annealing, the hardness of the sample remains basically unchanged. The results show that the CoCrFeNiTi$_{0.5}$ alloy has good microstructure stability and resistance to tempering and softening.

Jinxiong Hou et al. studied the effect of cold-rolling on the microstructure evolution and mechanical behavior of the Al$_{0.25}$CoCrFeNi high-entropy alloy [10]. Cold-rolling results in extensive grain elongation, formation of deformation bands within grains, and development of crystalline textures dependent on rolling reduction. The texture after cold-rolling was studied by electron backscattering diffraction. The current cold-rolled alloys have a strong brass texture. Cold-rolling results in a strong alloy. Its ultimate tensile strength is close to 1479 MPa, 2.8 times that of the as-cast state, but its plasticity is low (ε~2.3%).

Yong Zhang et al. systematically studied the alloying effects of Al, Ti, Cu, and Co in typical Ti$_x$CoCrFeNiCu$_{1-y}$Al$_y$ high-entropy alloys [11]. The phase transition of HEAs is closely related to the atomic packing efficiency of the phase. If the enthalpy of the alloy does not change much, the high entropy of the alloy will greatly reduce the Gibbs free energy, and the high-entropy phase can become an equilibrium stable phase. HEAs have ultrahigh fracture strength at room temperature and high temperature, such as a yield strength of 1400 MPa at 500°C.

W.H. Liu et al. studied CoCrFeNiMo$_x$ high-entropy alloys and found that face-centered cubic (FCC) high-entropy alloys (HEAs) showed excellent plasticity even at liquid nitrogen temperature, but their strength was relatively weak, which was far from meeting the requirements of practical structural applications [12]. One of the general concepts previously used in alloy design is the inhibition of the formation of brittle intermetallic compounds, which often results in severe embrittlement. Surprisingly, we revealed in this study that the precipitation of hard S and M intermetallic compounds greatly enhanced CoCrFeNiMo$_{0.3}$ HEA without causing severe embrittlement. Its tensile strength is up to 1.2 GPa, and its ductility is good, about 19%. It was found that the work-hardening index of the FCC matrix was as high as 0.75, which inhibited the growth of microcracks originating from these brittle particles.

Feng He et al. proposed a strategy for designing eutectic high-entropy alloys with high strength and ductility based on the existing binary phase diagrams of eutectic points [13]. A pseudo-eutectic binary alloy system of CoCrFeNiNb$_x$ (x = 0.1, 0.25, 0.5, and 0.8) was designed based on computer-aided thermodynamic calculation. The experimental results show that the eutectic alloy is composed of a ductile face-centered cubic (FCC) phase and a hard Laves phase with a good laminar flow structure. The designed alloy has excellent comprehensive mechanical properties of plasticity and strength. The compressive fracture strength and strain of the CoCrFeNiNb$_{0.5}$ alloy reached 2300 MPa and 23.6%, respectively.

Tao-tsung Shun et al. studied the effect of Ti addition on the microstructure and mechanical properties of the CoCrFeNiTi$_x$ (x = 0, 0.3, and 0.5) alloy [14]. The CoCrFeNi quaternary alloy exhibits a simple face-centered cubic solid-solution structure, while the mixed plate-like structure with (Ni, Ti) rich R phase and (Cr, Fe) rich S phase is observed in the face-centered cubic matrix of the CoCrFeNiTi$_{0.3}$ alloy.

In the CoCrFeNiTi$_{0.5}$ alloy, face-centered cubic matrix, rich (Ti, Co) Laves phase, and R-ship mixed phase are found. The compressive strength of the alloy increased by about 75% after Ti was added. The alloy with a high Ti content has higher yield stress and lower plasticity. Solution strengthening and secondary phase hardening of the face-centered cubic matrix are two main factors of alloy strengthening. The compressive strength and fracture strain of the CoCrFeNiTi$_{0.3}$ alloy are 1529 MPa and 60%, respectively. This indicates that this material shows the potential to develop into ductile high-strength alloys.

REFERENCES

[1] Tang Qunhua, Cheng Hu, Dai Pinqiang. Microstructure and Mechanical Properties of Al$_{0.3}$CoCrFeNi High Entropy Alloy after Rolling and Annealing [J]. *Journal of Materials and Heat Treatment*, 2015, 36(12):6.

[2] Sun Yongzhe, Tian Xiao, Wei Yafeng, et al. Effect of Heat Treatment on Microstructure and Mechanical Properties of Cold Rolled Al$_{0.3}$CoCrFeNi High Entropy Alloy [J]. *Journal of Materials and Heat Treatment*, 2017, 38(9):6.

[3] Si Songhua, Zhou Fangying, Wang Jianguo. Effect of Cold Rolling and Heat Treatment on Microstructure and Properties of Al$_{0.3}$CoCrFeNi High Entropy Alloy [J]. *Heat treatment of metals*, 2020, 45(3):6.

[4] Li D, Li C, Feng T, et al. High-entropy Al0.3CoCrFeNi alloy fibers with high tensile strength and ductility at ambient and cryogenic temperatures[J]. *Acta Materialia*, 2017, 123:285–294.

[5] Zhang Y, Zhang M, Li D, et al. Compositional design of soft magnetic high entropy alloys by minimizing magnetostriction coefficient in (Fe$_{0.3}$Co$_{0.5}$Ni$_{0.2}$)$_{100x}$(Al$_{1/3}$Si$_{2/3}$)$_x$ System[J]. *Metals*, 2019, 9(3):382.

[6] Zuo Tingting. Microstructure and Properties of Co-Fe-Ni High Entropy Magnetic Alloy [D]. University of Science and Technology Beijing, 2017.

[7] Shun T T, Chang L Y, Shiu M H. Microstructures and mechanical properties of multi-tiprincipal component CoCrFeNiTix alloys[J]. *Materials Characterization*, 2012, 70(OCT.30):63–67.

[8] Li Anmin, Wang Meihua, Shi Junzuo, et al. Effect of Forging on Microstructure and Properties of FeMnCrCoNi High Entropy Alloy [J]. *Forging & Stamping Technology*, 2019, 44(2):9.

[9] Jiang L, Lu Y, Dong Y, et al. Annealing effects on the microstructure and properties of bulk high-entropy CoCrFeNiTi$_{0.5}$ alloy casting ingot[J]. *Intermetallics*, 2014, 44:37–43.

[10] Hou J, Min Z, Ma S, et al. Strengthening in Al$_{0.25}$CoCrFeNi high-entropy alloys by cold rolling[J]. *Materials Science and Engineering: A*, 2017, 707(Nov. 7):593–601.

[11] Dean S W, Zhang Y, Chen G L, et al. Phase change and mechanical behaviors of TixCoCrFeNiCu1−yAly high entropy alloys[J]. *Journal of Astm International*, 2010, 7(5).

[12] Liu W H, Lu Z P, He J Y, et al. Ductile CoCrFeNiMox high entropy alloys strengthened by hard intermetallic phases[J]. *Acta Materialia*, 2016, 116:332–342.

[13] He F, Wang Z, et al. Designing eutectic high entropy alloys of CoCrFeNiNbx[J]. *Journal of Alloys & Compounds*, 2016, 656:284–289.

[14] Shun T T, Chang L Y, Shiu M H. Microstructures and mechanical properties of multi-principal component CoCrFeNiTix alloys[J]. *Materials Characterization*, 2012, 70(Oct. 30):63–67.

3 BCC-Structured High-Entropy Materials

Yong Zhang and Fangfei Liu

3.1 PRODUCTION METHODS OF BCC HEAs

There are a variety of methods that can be used in the production of HEAs [1–3], during which it is possible to carry out mass production with existing equipment and technologies, as no special equipment is needed [4]. HEAs can be easily fabricated into different forms, such as powders, thin films, and bulk materials [5].

According to the element mixing method in the preparation process of high-entropy alloys, the preparation process can be divided into three categories. The first is solid-state mixing, commonly mechanical alloying and subsequent consolidation. The second type is liquid-phase mixing, and the common ones are arc melting, induction melting, resistance melting, laser cladding, and laser near-net forming (LENS). The equipment involved in liquid-phase mixing preparation includes an electric arc furnace, induction furnace, magnetic levitation furnace, electron beam furnace, and single-crystal growth furnace. The third type is a gaseous mixture, commonly including sputtering deposition, vapor deposition, pulsed laser deposition (PLD), molecular beam epitaxy (MBE), and atomic layer deposition (ALD), through which films with a certain thickness can be prepared on the substrate.

3.1.1 SOLID-STATE MIXING

Mechanical alloying is a common method for the preparation of high-entropy alloys by solid-phase mixing. According to the proportion of principal components of the required high-entropy alloy, each principal component powder was added into the ball mill roller for mechanical mixing. In this process, the cold-welding phenomenon and the fracture of powder particles appear alternately, and finally, the atomic level alloying effect is achieved. The alloyed powder mixture can be formed into bulk by hot isostatic pressing sintering or discharge plasma sintering, and so on. The resulting bulk high-entropy alloy can also be processed, such as hot pressing or hot-rolling densification.

In 2008, Varalakshmi et al. [6] published the results of AlFeTiCrZnCu; the alloy powder with a single-phase BCC structure was formed by 20 h high-energy ball grinding. The powder grain size is about 10 nm. Subsequently, sintering using vacuum thermopressing resulted in a hardness of 9.50 GPa and alloy block of 2.19 GPa.

DOI: 10.1201/9781003319986-3

Using thermal isostatic pressure for sintering, prepare alloy blocks with a hardness of 10.04 GPa and a compressive strength of 2.83 GPa. The results show that hot isostatic pressure sintered alloy blocks are more compact, with better mechanical properties.

Mechanical alloying (MA) with spark plasma sintering (SPS) [7–9], as a typical technology of PM, can readily fabricate bulk high-density HEAs with ultrafine grains, excellent microstructural homogeneity, and improved strength and hardness. $Nb_{25}Mo_{25}Ta_{25}W_{25}$ and $Ti_8Nb_{23}Mo_{23}Ta_{23}W_{23}$ HEAs [10] were successfully prepared by MA with SPS technology. The compressive yield stress and fracture strain of $Nb_{25}Mo_{25}Ta_{25}W_{25}$ HEAs with average grain sizes ~0.88 μm are 2460 MPa and 16.8%, respectively, which are remarkably superior to those prepared by casting [11].

Long et al. [12] reported that mechanically alloyed (MA) NbMoTaWVCr refractory high-entropy alloy (HEA) powders were sintered by spark plasma sintering (SPS) at temperatures of 1400°C–1700°C. They discussed that during the MA process, only a supersaturated body-centered cubic (BCC) structured solid solution was formed. When the sintering temperature increased to 1600°C, the Laves phase was transferred to a C_{14} structure, and its volume fraction was dramatically reduced. The plasticity of the refractory HEA was strongly affected by the fraction, size, and distribution of the Laves phase and oxide particles. The NbMoTaWVCr alloy sintered at 1500°C obtained an excellent combination of yield strength (3416 MPa) and failure plasticity (5.3%) at room temperature.

Cao et al. [13] successfully prepared $TiNbTa_{0.5}Zr$ and $TiNbTa_{0.5}ZrAl_{0.2}$ HEAs with a single BCC phase using powder metallurgy technology. The compressive yield strength and strain for $TiNbTa_{0.5}Zr$ and $TiNbTa_{0.5}ZrAl_{0.2}$ alloys at room temperature were 1310 MPa, 30%, and 1500 MPa, 30%, respectively.

In addition to the samples earlier, BCC-structured HEA systems prepared by MA with SPS technology also include AlFeTiCrZnCu [14–16], CrTiVTaW [17,18], TiNbTaZr [19], and FeCrMnV [20–22]. The AlFeTiCrZnCu HEA [16] prepared by MA with SPS technology can achieve a 99% density and homogeneous nanostructure (~10 nm), and its hardness can reach 2 GPa.

In general, most BCC HEAs follow the strength-ductility trade-off; however, $TiNbTa_{0.5}Zr$ and $TiNbTa_{0.5}ZrAl_{0.2}$, prepared by powder metallurgy by Cao et al., present a good combination of strength and plasticity, while most refractory HEAs still follow the strength-ductility trade-off. Besides this, $TiNbTa_{0.5}Zr$ and $TiNbTa_{0.5}ZrAl_{0.2}$ HEAs show a compressed maximum engineering strain of 50% without any cracking or fractures at 800 °C.

The volume fraction of the BCC2 phase gradually increased with an increase in the Al concentration in $Al_xCrFeMoV$ HEAs [23]. The improvement of compressive yield strength from 2730 to 3552 MPa can be attributed to the solid-solution strengthening of Al caused by the appearance of the BCC2 phase. The addition of Al dually influenced the properties of the CrFeMoV alloy by improving its strength and reducing the density of the system. The yield strength and hardness as a function of density were compared with data for previously reported HEAs. $Al_xCrFeMoV$ HEAs with outstanding mechanical properties, low cost, and low density, which are better than those of any previously reported HEAs, suggested a promising future for HEAs in many structural applications.

In another case, new phases appeared in the BCC matrix, such as the B2 phase, HCP phase, and FCC phase, after SPS processing. As reported [24], AlCuFeMnTiV HEA prepared by sintering powder containing only the BCC phase has the B2 phase, HCP phase, and Cu-rich FCC phase precipitated at the grain boundary in addition to the BCC matrix. It exhibits the best comprehensive mechanical properties, with a density of 6.28 g/cm^3, compressive yield strength of 2060 MPa, and plastic strain of 15.83%, which are superior to most LWHEAs and traditional lightweight alloys. The high strength and good plasticity of AlCuFeMnTiV HEAs are attributed to the strengthening effect of nano-twins precipitated in the FCC phase on grain boundaries.

Powder metallurgy (PM), a forming technology that allows significant compositional accuracy, can completely prevent chemical segregation, can obtain a homogeneous microstructure, can produce nanocrystalline materials, and can develop metal matrix composites. Lightweight RHEAs of CrNbVMo exhibited superior compressive specific yield strengths compared to cast RHEAs at 25°C and 1000°C [25]. TiNbTa$_{0.5}$Zr and TiNbTa$_{0.5}$ZrAl$_{0.2}$ RHEAs were successfully prepared without any cracks or fractures by PM, and with Al addition, the compressive peak stress increased from 508 MPa to 603 MPa at 800°C [26]. The PM process is not only suitable for preparing RHEAs but is also suitable for preparing small and precise components.

The earlier discussion demonstrates that powder metallurgy is a promising way of preparing ductile RHEAs with outstanding comprehensive mechanical properties.

Universal property is a significant feature of mechanical alloying, and almost all materials can be prepared by this method, such as tough metal alloys, brittle intermetallic compounds, and composites. Mechanical alloying can be carried out at room temperature, so the emergence of chemical precipitation in melting and casting is avoided, and the problem of difficult solidification processing due to different melting temperatures or air pressures is also solved. However, during ball milling, the alloy powder may form a new phase by mixing with impurities, thus affecting the alloy properties. Impurities may come from ball mill media, process control agents, and the atmosphere in the ball grinding tank. Therefore, an appropriate ball grinding medium and a process control agent should be selected for mechanical alloy preparation of high-entropy alloys, and in order to prevent oxidation or nitride, the process can be protected under vacuum or by an inert gas.

3.1.2 Liquid-State Mixing

Liquid-state mixing methods include arc melting and laser cladding. Arc smelting is widely used in HEAs production. In arc smelting, liquid materials are mixed in a furnace. Disadvantages of arc melting It is not easy to control the rapid solidification process. The limitation of laser cladding method is that it can only be treated in a small area. In addition, it is necessary to prevent fracture and porosity in order to obtain good organization and performance [27].

In addition to vacuum arc melting method, vacuum induction method is also used, which is the most widely used HEA production method. Vacuum arc melting is a popular method because the temperature released during arc melting (about 3000° C) is sufficient to melt many metals used in HEA production [28].

Arc melting process is to achieve production by melting various elements in an arc melting furnace. The temperature of the arc melting furnace can reach 3000°C. This can be controlled by adjusting electrical parameters. Therefore, most high melting point elements can be used in this way. This method is suitable for elements with high melting point. If there are low melting point elements (such as Mg, Zn, Mn) in the composition, it is difficult to achieve content control [29].

Another limitation of this methods is the appearance of a heterogeneous structure with different separation mechanisms due to low solidification. The dendritic and interdendritic structures in the typical solidification microstructure encountered in the production of HEAs using the vacuum arc technique.

Arc melting is the most commonly used method in the preparation of high-entropy alloy. Repeat melting many times so that each principal element is fully mixed in the liquid state to ensure the acquisition of an alloy with uniform chemical composition and then pour into the copper mold for cooling and solidification molding. When the melting temperature is low, the induction melting method can also be used. Liquid mixing should be carried out under vacuum and other protective atmospheres to avoid the mixing of oxygen and nitrogen into the excess phase so as to ensure the mechanical properties and purity of the alloy.

Vacuum arc melting adopts a vacuum arc furnace, which can be divided into two categories according to the electrode materials: self-consuming electrode and non-self-consuming electrode. The self-consuming electrode uses the elements of each main element or the high-entropy alloy rod as the material. The electrode discharges heat under vacuum conditions, and the electrode is melted and mixed to obtain the alloy with uniform composition. Non-self-consuming electrode with tungsten rod as material, through the power-induced arc, after main element raw material heating and smelting, repeated melting can ensure that the alloy chemical composition is uniform.

The vacuum induction melting method is necessary to adjust the loading order of each main element. In order to ensure the complete melting of the high melting point raw materials, the low melting point element should be placed at the bottom, and the high melting point element should be put in later. In addition, in order to avoid refractory compound generation, it is also necessary to separate the raw materials prone to compound reactions. Electromaglev coil can be used to avoid contamination with crucible material.

After the smelting, the homogeneous liquid alloy can be poured into the copper mold, and the alloy block can be obtained. The copper mold has a strong heat dissipation capacity, which can ensure that the tissue grain is small and avoid uneven composition due to different diffusion capacities and other factors. Other solidification methods can also be adopted according to the requirements, such as the directional solidification method, which will ensure the growth direction of the microstructure. The obtained alloy block has a certain crystal orientation to achieve the purpose of controlling the organization and optimizing the performance.

In contrast to the conventional process, additive manufacturing (AM) is based on an incremental layer-by-layer fabrication process [30]. Because local process control can be realized in the AM process, it has extremely rapid solidification cooling rates, a production with unrivaled design freedom, and a shorter production cycle [31,32]. The AM process can overcome the inherent complexity and achieve the high levels of control required to produce homogeneous bulk alloys. Due to the immanent advantages of AM, it has attracted much attention in the last decade. At present, the process for preparing HEAs mainly includes direct laser deposition (DLD), selective laser melting (SLM), and selective electron beam melting (SEBM) [33–36].

Recently, several attempts have been made to prepare BCC-structured HEAs (mainly RHEAs) by laser deposition techniques, and there have been many studies on the preparation of those HEAs by arc melting and PM. The attempts to prepare BCC-structured HEAs using AM technology have mainly been based on DLD technology, and few reports have been published on SLM or SEBM.

The process of DLD may also be known as laser metal deposition (LMD), direct metal deposition (DMD), laser-engineered net shaping (LENS), and laser cladding. The MoNbTaW RHEA [37] with single-wall structures was the first BCC-structured alloy system to be prepared by the DLD method. The case indicated that it is feasible to fabricate a RHEA through in situ alloying of a MoNbTaW elemental mixture even if cracks appear during processing. The bulk of the crack-free TiZrNbTa RHEAs [38] was successfully produced by in situ alloying of elemental powders using the DLD method. It is one of the few alloy systems with suitable room-temperature plasticity among the RHEAs. A well-defined compositional gradient with good hardness (440 HV0.1) was obtained by optimizing the DLD method.

SLM, as a typical AM technology, can three-dimensionally (3D) fabricate components with intricate shapes and refined resolutions [39–41]. The fabrication of BCC-structured HEAs through the SLM process has rarely been considered. A MoNbTaW refractory HEA was prepared via the SLM process using blended elemental powders [36,42]. There was a deviation between the chemical composition of the prepared sample and that of a pre-mixed powder. This was likely due to surface evaporation of the lower melting point elements, which floated to the upper surface of the melt pool during the SLM process. Composition partitioning was in direct contrast with that reported by Dobbelstein et al. [15] for MoNbTaW HEAs fabricated via DLD.

The operating principle was similar to that of SLM, but SEBM used an electron beam instead of a laser beam as the heat source, which meant it had attracted extensive attention in recent years. SEBM has the unique characteristics of a high energy density of the incident electron beam, high scan speed, moderate operation cost, and so on. In addition, the high-temperature preheating (up to 1100°C) of the powder bed by the electron beam prior to scanning and melting is another distinct working condition of SEBM. This working condition results in low residual stress of the built products, making the SEBM process suitable for fabricating complex-shaped products and reducing the thermal cracking and distortion of the printed HEAs [43–45].

Hiroshi et al. [44] investigated the microstructures and mechanical properties of equiatomic AlCoCrFeNi HEA samples fabricated by SEBM, comparing them with those of samples prepared by arc melting. The proportion of the FCC phase (pre-cipitated at the grain boundaries of the B2/BCC grains) for the bottom was much higher than that for the top. Therefore, the hardness of the SEBM samples gradu-ally decreased with an increased proportion of the FCC phase, and they exhibited much higher plastic deformability than the cast specimen without a significant loss of strength.

Wang et al. precipitated CoCrFeNiMn HEA via EBM; the tensile properties of which (average YS ~205 MPa, elongation ~63%) were almost the same as those of the as-cast form obtained by He et al. [46]. The $Al_{0.5}CrMoNbTa_{0.5}$ HEAs [34] from elemental powder blends were prepared using the SEBM technique. By optimizing the process parameters, the porosity was reduced to replace the post-treatment in the traditional processing technology. However, the simultaneous handling of several elemental or pre-alloyed powders brings new challenges to the deposition process.

3.1.3 GAS-STATE MIXING

The gas-state mixing method includes physical vapor deposition (PVD). With this method, thin HEA film coatings can be obtained [47,48]. Magnetron sputtering depo-sition is the most popular method used to produce HEA films. Sputtered atoms close to the atomic fractions of the deposited HEA are thrown from the sputtering gas by atomic or ion bombardment. The sprayed atoms are randomly deposited on the sur-face of the substrate, and it is made into a HEA film and grows. Therefore, the micro structure of the HEA film is determined by many factors, including target materials, power, atmosphere, base pressure and temperature [48].

The gas-state mixing methods are mainly used to prepare high-entropy alloy films, among which physical vapor deposition (PVD) is the most common. Vacuum magnetron sputtering is a kind of PVD, with high-entropy alloy or each main block as the target, controlling the energy exchange of the target through the magnetic field; the target surface metal flies out in an atomic or ionic state, and adsorbed and deposited on the substrate, forming a thickness of a certain film. According to the sputtering mode, it can be divided into reaction magnetron sputtering, pulse magne-tron sputtering, and equilibrium and nonequilibrium magnetron sputtering. During the sputtering process, the raw material form, atmosphere, temperature, and pressure of the base will all affect the shape, growth, and microstructure of the high-entropy alloy film.

When preparing the high-entropy alloy film by using chemical vapor deposi-tion, it is necessary to place the compounds containing the principal elements of the high-entropy alloy in the reaction atmosphere and produce the gas forming the high-entropy alloy elements through thermal volatilization and chemical reaction, and then the gas is adsorbed and deposited on the substrate to form the film. Because it is difficult to control the proportion of each element in the produced gas, it is dif-ficult to prepare a high-entropy alloy film conforming to the chemical composition. Atomic layer deposition controls the atomic layer and film growth through continu-ous surface reaction. Due to a large number of high-entropy alloy team members, film preparation using this method requires multiple atomic layer deposition cycles. Molecular beam epitaxy technology is widely used in film preparation. Due to the high requirements of vacuum during its preparation and the very slow deposition rate, the high purity and high-entropy alloy film can be prepared, and the film is often in a single-crystal state. The development of high-entropy thin-film technology enables the high-throughput preparation of high-entropy alloys. High-entropy alloy films with a composition gradient by magnetron sputtering were prepared by Zhang et al. The three targets of Al target, Ti target, and CrFeNi ternary target were, respectively, 120 apart and focused on the silicon wafer with a silicon oxide coating. Using the spatial relationship between the three targets and the substrate, the film elements were gradient changed, and 144 independent sample units with different compositions were obtained at one time. Subsequent characterization of high-entropy thin films revealed that samples in both Al-rich and Ti-rich regions were more prone to form more rigid BCC structures.

3.2 MECHANICAL BEHAVIOR OF BCC HEAs

HEAs in the first category consist of many refractory elements, so-called refractory high-entropy alloys (RHEAs), and mostly form BCC structures. In contrast to FCC HEAs, BCC HEAs exhibit relatively high intrinsic yield strengths. Besides, RHEA systems have excellent high-temperature mechanical properties but insufficient toughness at room temperature.

3.2.1 HARDNESS AND COMPRESSION BEHAVIOR OF BCC HEAs

Due to the permanent deformation, Vickers hardness (HV) correlates with the σy according to the following empirical equation [49]:

$$HV \sim (2.5 \text{ to } 3.0) \times \sigma y$$

The Vickers hardness of HEAs is strongly dependent on the crystal structure [50]. For example, He et al. [51] found that the hardness values of the $(FeCoNiCrMn)_{100-x}Al_x$ HEA increase slightly from 176 HV for the FeCoNiCrMn HEA to 182 HV for the $(FeCoNiCrMn)_{93}Al_7$ HEA (all alloys are FCC structured). With the increase of the Al content, the structure of HEAs turns into a mixture of FCC + BCC phases, and the hardness value significantly increases.

Near-equiatomic WNbMoTa with a single-phase BCC structure was first proposed by Senkov [52], where the cast sample density and Vickers microhardness were $\rho = 13.75$ g/cm^3 and $Hv = 4455$ MPa, respectively. The alloy possessed a compressive yielded strength of 1058 MPa [53] failed by splitting at $\varepsilon_p = 2.1\%$ at an ambient temperature and decreased in yielded strength to 561 MPa and 552 MPa when the samples deformed at 600°C and 800°C, respectively [54].

After compression deformation, scanning electron microscopy (SEM) imaging revealed that the fracture morphologies of the alloys contained a brittle quasi-cleavage fracture, indicating that the primary failure mode in this alloy is tensile rather than shear.

For another refractory HEAs, Senkov et al. [55] reported that the σy of the as-cast NbMoTaW HEA was ~1,058 MPa, but its relatively low compression plasticity (only 1.5%) prevents its further application. The microstructure of the NbMoTaW HEA was BCC structure with a small number of segregations (< 5%) at the dendritic grain boundary. The fracture morphology indicates that the failure mode was with longitudinal crack rather than shear cracks, which corresponds to the brittle quasi-cleavage fracture.

Yao et al. [56] calculated and designed NbTaTiV, NbTaVW, and NbTaTiVW with a single phase, and the results showed that NbTaTiV exhibits exceptional compressive ductility (~50%) at room temperature and yield strength of 965 MPa, while NbTaTiVW and NbTaVW show yield strengths of 1420 MPa and 1530 MPa with fracture strains of 20% and 12%, respectively. Work hardening can be observed in these HEAs.

For instance, the AlCrFeCoNi HEA with a single-phase BCC solid solution exhibits excellent compressive properties of yield stress (1250.96 MPa) and plastic strain (32.7%) [57]. VNbMoTa RHEAs exhibit excellent room-temperature ductility with a fracture strain > 25% and high-temperature strength (compressive yield strength of 811 MPa at 1000°C) [58]. However, the cast product has some drawbacks, including many structural defects, such as voids, porosity, chemical segregation, and grain coarsening.

3.2.2 Tensile Behavior of BCC HEAs

Some HEAs with BCC structure also exhibit some tensile plasticity behavior, such as the TaNbHfZrTi HEA. Senkov et al. [59] examined the tensile properties of the TaNbHfZrTi HEA after cold-rolling and then annealing at 1073 K and 1273 K, respectively. The σy of the annealing TaNbHfZrTi HEA (cold-rolled plus annealed at 1073 K) can exceed ~1,100 MPa, and the elongation at break was able to exceed ~9% at room temperature.

Lilensten et al. [60] studied the underlying deformation mechanism of the TaNbHfZrTi HEA by SEM, electron backscattering diffraction (EBSD), and TEM. When being strained to 2.65%, active dislocation slips with a planar slip mode were present, and no deformation twinning was observed. Moreover, the activities of dislocations were evident at grain boundaries.

The TEM images of the TaNbHfZrTi HEA prestrained to different strains. When the plastic deformation initiates, there are two slip systems that correspond to <111> screw dislocations. And there are also some dislocation dipoles and loops. When the strain is higher, the two slip systems can form dislocation bands and dislocation-free zones. With an increase in the strain, the distance between the bands decreases, as exhibited in Figures 19(c) and (d). These results indicate that screw dislocations are present during the entire deformation process.

3.3 STRENGTHENING OF BCC HEAs

In HEAs, the concepts of solute and solvent are no longer applicable in conventional alloys. On this basis, the reinforcement theory also needs to be tested. Therefore, the research in this field is a hot spot. The research in this field is a hot topic of HEAs. In recent years, scientists have done a great amount of work on how to strengthen HEAs. It is worth noting that the deformation behavior of metals is closely related to their defects contents, such as dislocations and twins [61,62]. Therefore, it is of great importance to manipulate the defects to strengthen the HEAs. Here, the traditional strengthening methods are discussed, such as grain-boundary strengthening, solid-solution strengthening, and particle strengthening.

3.3.1 GRAIN-BOUNDARY STRENGTHENING

Juan et al. [63] refined that the grain size of the BCC-structured TaNbHfZrTi HEA by controlling the annealing temperature and time, which effectively improved its strength and plasticity. It has been reported that the Nb and Ta elements could affect the growth rate of grain, resulting in slower grain boundary migration. Thus, the solute resistance effect produces an enhancement of HEA.

3.3.2 SOLID-SOLUTION STRENGTHENING

The principle of solid solution strengthening is to dissolve alloying elements into the matrix material, causing a certain degree of lattice deformation, thus improving the strength of the alloy. Solid solution reinforcement can be divided into two types: substitution solid solution reinforcement and interstitial solid solution reinforcement.

Studies have shown that the effect of substituting solid solution reinforcement on these types of HEAs is very limited. Interstitial solution strengthening refers to the elements dissolved in the interlattice position of the matrix, usually matrix atoms with radius difference of more than 41%, such as H, C, B, N, O, etc. [64–68]. Compared with substitutional atoms, interstitial atoms can interact more with the defects in HEAs so that the strengthening effect is more significant [69,70]. In recent years, due to the combination of this gap strengthening effect and relatively low manufacturing cost, HEAs doped with gap elements have attracted more attention from scientists [69–81].

Chen et al. [82] added 0.05, 0.1, and 0.2 at.% of oxygen to the $ZrTiHfNb_{0.5}Ta_{0.5}$ refractory HEA to improve the strength of the alloy. Lei et al. [83] added 2.0 at.% O to the TiZrHfNb HEA. The results showed that UTS increased by 48.5% and ductility increased by 95.2%, which solved the strength-ductility tradeoff problem.

3.4 PARTICLE STRENGTHENING

3.4.1 TRIP EFFECT

TRIP effects have been shown to significantly improve both properties and plasticity in conventional materials. For fcc alloys, SFE can significantly affect the deformation mode. When the SFE value is high, the deformation mode is dislocation slip. For SFE, twin is the main deformation mode. When the SFE value is low, the deformation mode is phase transition. For alloys with low SFE value, TRIP effect is very obvious. TRIP effect is also used to design HEAs with excellent comprehensive mechanical properties.

Huang et al. [84] applied TRIP effect to TaNbHfZrTi fire resistance HEA. By reducing the content of Ta in HEA, the stability of bcc phase is reduced, and the two-phase microstructure of bcc phase and hcp phase is obtained. Therefore, when external forces are applied, the bcc phase changes to the hcp phase. Although the strength of TaNbHfZrTi HEA decreased with the decrease of Ta content, the ductility increased significantly with the introduction of TRIP effect, and a significant work hardening effect was produced.

3.4.2 Eutectic HEAs

The eutectic HEAs offer a possibility to solve the strength-ductility tradeoff, thus simplifying the industrial production of HeAs. Since the eutectic reaction is an isothermal transition process, there is no range of solidification temperature that reduces the content of both separation and shrinkage pores.

3.4.3 Strengthening by Fine Particles

Fine particles whose average sizes are below 1 micrometer and can distribute uniformly in the matrix, resulting in a strong inhibition of the dislocation slip. The strengthening effect can be determined by the radii of the particles (r) and the mean spacing between the particles (2λ) on the chosen slip plane.

3.5   SPECIAL PERFORMANCE AND APPLICATION OF BCC HEAs

3.5.1   Mechanical Behaviors of BCC HEAs at High Temperatures

The BCC high-entropy alloys have received much attention due to their excellent high-temperature oxidation properties and have natural advantages over the high-entropy alloys of other structures.

Ti elements and their alloys possess good ductility and excellent high-temperature properties. For example, the solid-solution hardening effect of Ti addition is beneficial to the compressive strength and ductility of NbMoTaW and VNbMoTaW HEAs at room temperature. The yield strengths of TiNbMoTaW and TiVNbMoTaW are ~586 and ~659 MPa at 1200°C, respectively [85], so they are expected to be used as materials for high-temperature applications.

The strength of TaNbHfZrTi HEA at high temperature is significantly reduced compared to the strength at room temperature [86]. When the temperature is lower than 873 K, the main deformation mode of TaNbHfZrTi HEA is dislocation slip, supplemented by a small amount of deformation twinning. Because the deformation process is uniform and continuous, its strength changes little. At temperatures above 1073 K, the higher diffusion rate promotes the crystallization of grain boundaries. In this case, the crack spreads rapidly. As a result, the strength of the TaNbHfZrTi HEA dropped dramatically from 535 MPa at 1073 K to 295 MPa at 1273 K. NbMoTaW HEA, another refractory alloy with bcc structure, exhibits brittle-ductile behavior at higher temperatures than the ductile - brittle transition temperature. This behavior indicates that the plasticity of the material has been improved. Based on the above discussion, refractory heat has natural advantages, making it a potential candidate for use in ultra-high temperature environments. [87]. The NbMoTaW HEA, which is another kind of the BCC-structured refractory alloy, exhibits brittle to ductile behavior [68]. The direction of cracking in the alloy is approximately 40° from that of the compression direction, while the fracture of the material was completed by a shear. Such behavior indicates that the plasticity of the material was improved [68]. Based on the earlier discussion, refractory HEAs possess natural advantages that make them potential candidates for applications in ultrahigh-temperature environments (873 K to 1873 K).

3.5.2 SELF-SHARPENING

The previous three parts of the article mainly summarized the BCC-structured HEAs prepared by three forming methods, largely focusing on their tensile or compressive properties under quasi-static conditions, while it is dynamic tensile or compressive properties that affect its self-sharpening.

"Self-sharpening" is the ability of a material to maintain its acute head shape during penetration, which is a necessary property of material during armor-piercing [88]. HEAs possess a good combination of strength and ductility, which is the premise of excellent self-sharpening. Besides this, high susceptibility to adiabatic shear banding (ASB) is the fundamental cause of self-sharpening behavior that benefits the penetration performance. ASB is the dominant deformation mechanism for materials consisting of metals or alloys under high-strain-rate loading, which exhibits a narrow band where large shear deformation occurs in a very short time.

At present, the research on self-sharpening mainly focuses on W-based alloys [89,90]. W-based alloys are promising candidates for kinetic energy penetrators because of their high density, strength, and ductility. However, the low susceptibility of W-based alloys to ASB reduces their penetration depth. Thus, it is necessary to develop a new matrix material to replace W-based alloys. The sluggish diffusion and lattice distortion effect of HEAs can achieve a balance between strength and toughness and improve penetration ability.

This means that HEAs can be used as potential materials with excellent self-sharpening properties. However, only limited success has been achieved in the development of new tungsten HEAs. A new chemical-disordered multi-phase tungsten HEA (WFeNiMo) was developed by Liu et al. [91]. Compared with conventional W-based alloys, WFeNiMo consists of a BCC dendrite phase and a rhombohedral μ phase embedded in the continuous FCC matrix, which means it exhibits outstanding self-sharpening capability.

This is due to the precipitation of the ultrastrong μ phase of WFeNiMo, which can mediate the shear banding by triggering dynamic recrystallization softening. Subsequently, Chen et al. [92] conducted experiments with WFeNiMo HEA and W-based alloy projectiles penetrating medium-carbon steel by using a ballistic gun and a two-stage light-gas gun. As the impact velocity increased (1330 m/s ~ 1531 m/s.), the penetration mode of the WFeNiMo HEA projectile changed from self-sharpening to mushrooming.

In general, BCC-structured HEAs have ultrahigh strength, high hardness, and room-temperature plasticity, which has been gradually improved in recent studies. It has great potential in aerospace, military (high-performance penetrator materials), and biomedical fields in future applications.

3.5.3 OXIDATION

As the main part of BCC-structured HEAs, RHEAs are promising candidates for new-generation high-temperature materials. Yet RHEAs have two fatal disadvantages: room-temperature brittleness and unsatisfactory high-temperature oxidation resistance. Early research was on improving the high-temperature oxidation resistance of RHEAs by adding elements such as Al, Cr, and Si. However, this tends to promote the formation of intermetallic compounds, such as Al_2O_3, Cr_2O_3, and SiO_2, which worsen the poor room-temperature ductility. In this context, it presents a challenge to enhance the oxidation resistance and achieve optimal room-temperature ductility at the same time for RHEAs.

Müller et al. [93] systematically investigated the oxidation behavior of four refractory HEAs within the system of the TaNbMoCrTiAl RHEA (including TaMoCrTiAl, NbMoCrTiAl, NbMoCrAl, and TaMoCrAl) at 900 and 1100°C in the air. Because of the formation of protective Al_2O_3, Cr_2O_3, and $CrTaO_4$ oxide layers, the TaMoCrTiAl RHEA shows superior oxidation resistance at 1000°C in the air. Moreover, the addition of Ti can effectively improve the oxidation resistance of the RHEA, which is attributed to promoting the formation of protective rutile-type oxides and reducing the formation of less favorable oxides. However, the oxidation resistance of the RHEAs is significantly reduced with V addition. It is mainly due to the addition of V that makes the oxide scales become porous for the other RHEAs and aggravates the volatility of V_2O_5, and leads to disastrous internal oxidation.

Those results add a crucial perspective to the further development of RHEAs as novel high-temperature materials with balanced room-temperature ductility and high-temperature oxidation resistance.

3.5.4 CORROSION

HEAs exhibit much better corrosion resistance than traditional corrosion-resistant metal materials (e.g., stainless steel, copper-nickel alloys, and high-nickel alloys), which is attributed to the high-entropy and cocktail effects, in particular. Therefore, the corrosion resistance of HEAs has attracted extensive attention in the field of corrosion research.

Qiu et al. [94] reviewed the influence of metal elements (including aluminum, titanium, chromium, molybdenum, and nickel) and processing methods (anodizing and aging) on the corrosion resistance of HEAs. $AlMo_{0.5}NbTa_{0.5}TiZr$ HEA [95] exhibited extensive segregation of alloying elements and significant gradients in local chemistry, which lend it excellent corrosion characteristics. Fu et al. [96] further reviewed the corrosion behaviors and mechanisms of HEAs in various aqueous solutions; discussed the effects of heat treatment, anodizing treatment, and preparation methods on the corrosion behaviors of HEAs; and established correlations between the composition, microstructure, and corrosion resistance of HEAs.

The electrochemical behaviors of the AlCoCrFeNi HEA obtained with SEBM were investigated, which were influenced by the phase morphologies. The pitting potential of SEBM specimens (0.112 V vs. Ag/AgCl) was lower than that of a cast specimen (0.178 V vs. Ag/AgCl). Equiatomic TiZrNbTaMo RHEA [97] as cast samples underwent a study of their corrosion resistance to investigate whether they met the requirements of orthopedic implants. Compared with Ti_6Al_4V, 316LSS, and CoCrMo alloys, TiZrNbTaMo HEA exhibited excellent corrosion resistance in a phosphate buffer solution (PBS) and potentially excellent biocompatibility, attributed to the surface passivation and high stability, regardless of the pitting. It has preliminary advantages in mechanical properties and corrosion resistance and can offer an opportunity to explore new orthopedic-implant alloys.

3.5.5 IRRADIATION

Under the condition of irradiation, structural materials are faced with the interaction between irradiated particles (ions, neutrons, electrons, etc.) and lattice atoms of the material itself, which lead to the formation of irradiation defects and the evolution of microstructure, and then affect the service performance of the material. Among them, the most difficult problem and challenge is volume swelling. HEAs have the advantage of being an irradiation-resistant material because of their high-temperature phase structure stability, high-temperature softening resistance, high-temperature oxidation performance, and corrosion resistance comparable to that of austenitic stainless steel.

As a general rule, BCC alloys exhibit superior resistance to void swelling, primary defect production, and defect evolution behaviors than FCC alloys. Compared with FCC alloys, the peak in BCC alloys swelling is often found at comparatively low homologous temperatures when they do undergo void swelling and their temperature dependence of swelling can be quite different. Because of higher atomic diffusivities, BCC alloys tend to have a higher speed in the diffusion-dependent process. Therefore, BCC-structured HEAs have attracted extensive attention because of their unique high-temperature stability and radiation resistance.

A BCC-structured W-based RHEA film with outstanding radiation resistance has been prepared by O. El-Atwani [98]. In their work, the films with uniform composition have element segregation at the grain boundary after irradiation, and the precipitation of Cr-rich and V-rich phases occurs in the grains. The hardness of the deposited film is about 14 GPa, and the hardness polarization does not change after heat treatment and irradiation, indicating that the alloy has excellent radiation-softening resistance and can maintain the stability of the microstructure.

Two novel BCC-structured $Mo_{0.5}NbTiVCr_{0.25}$ and $Mo_{0.5}NbTiV_{0.5}Zr_{0.25}$ HEAs were fabricated by vacuum arc melting [99]. Zhang et al. investigated the crystal structure, hardness, and microstructure evolution performed on the two HEAs to simulate neutron irradiation with Helium-ion irradiation. The two HEAs showed slight irradiation hardening compared with most of the conventional alloys. The helium bubbles and dislocation loops with small sizes were observed in the two HEAs after irradiation. This is the first report on the formation of a dislocation loop in BCC-structured HEAs after irradiation. $Mo_{0.5}NbTiVCr_{0.25}$ and $Mo_{0.5}NbTiV_{0.5}Zr_{0.25}$ HEAs show outstanding irradiation resistance, which may be promising accident-tolerant fuel-cladding materials.

REFERENCES

[1]   Ma, S.; Zhang, S.; Gao, M.; Liaw, P.; Zhang, Y. A successful synthesis of the CoCrFeNiAl0.3 single-crystal, high-entropy alloy by Bridgman solidification. *JOM*, 2013, 65(12), 1751–1758.

[2]   Ye, X.; Ma, M.; Liu, W.; Li, L.; Zhong, M.; Liu, Y., et al. Synthesis and characterization of high-entropy alloy Al. *Advances in Materials Science and Engineering*, 2011, 2011.

[3]   Novak, T.G.; Vora, H.D.; Mishra, R.S.; Young, M.L.; Dahotre, N.B. Synthesis of Al 0.5 CoCrCuFeNi and Al0.5CoCrFeMnNi high-entropy alloys by laser melting. *Metallurgical and Materials Transactions B*, 2014, 45(5), 1603–1607.

[4]   Tsai, M.H.; Yeh, J.W. High-entropy alloys: A critical review. *Materials Research Letters*, 2014, 2(3), 107–123, http://doi.org/10.1080/21663831.2014.912690.

[5]   Jiang, Z.; Chen, W.; Xia, Z.; Xiong, W.; Fu, Z. Influence of synthesis method on microstructure and mechanical behavior of Co-free AlCrFeNi medium-entropy alloy. *Intermetallics*, 2019, 108, 45–54.

[6]   Varalakshmi, S.; Kamaraj, M.; Murty, B.S. Synthesis and characterization of nanocrystalline AlFeTiCrZnCu high entropy solid solution by mechanical alloying[J]. *Journal of Alloys and Compounds*, 2018, 460(1/2), 253–257.

[7]   Joo, S.-H.; Kato, H.; Jang, M.J.; Moon, J.; Kim, E.B.; Hong, S.-J.; Kim, H.S. Structure and properties of ultrafine-grained CoCrFeMnNi high-entropy alloys produced by mechanical alloying and spark plasma sintering. *Journal of Alloys and Compounds*, 2017, 698, 591–604, http://doi.org/10.1016/j.jallcom.2016.12.010.

[8]   Sathiyamoorthi, P.; Basu, J.; Kashyap, S.; Pradeep, K.G.; Kottada, R.S. Thermal stability and grain boundary strengthening in ultrafine-grained CoCrFeNi high entropy alloy composite. *Materials & Design*, 2017, 134, 426–433, http://doi.org/10.1016/j.matdes.2017.08.053.

[9]   Wang, P.; Cai, H.; Zhou, S.; Xu, L. Processing, microstructure and properties of Ni1.5CoCuFeCr0.5–xVx high entropy alloys with carbon introduced from process control agent. *Journal of Alloys and Compounds*, 2017, 695, 462–475, http://doi.org/10.1016/j.jallcom.2016.10.288.

[10] Pan, J.; Dai, T.; Lu, T.; Ni, X.; Dai, J.; Li, M. Microstructure and mechanical properties of Nb25Mo25Ta25W25 and Ti8Nb23Mo23Ta23W23 high entropy alloys prepared by mechanical alloying and spark plasma sintering. *Materials Science and Engineering: A*, 2018, 738, 362–366, http://doi.org/10.1016/j.msea.2018.09.089.

[11] Senkov, O.N.; Wilks, G.B.; Scott, J.M.; Miracle, D.B. Mechanical properties of Nb25Mo25Ta25W25 and V20Nb20Mo20Ta20W20 refractory high entropy alloys. *Intermetallics*, 2011, 19, 698–706, http://doi.org/10.1016/j.intermet.2011.01.004.

[12] Long, Y.; Liang, X.; Su, K.; Peng, H.; Li, X. A fine-grained NbMoTaWVCr refractory high-entropy alloy with ultra-high strength: Microstructural evolution and mechanical properties. *Journal of Alloys and Compounds*, 2019, 780, 607–617, http://doi.org/10.1016/j.jallcom.2018.11.318.

[13] Cao, Y.; Liu, Y.; Liu, B.; Zhang, W. Precipitation behavior during hot deformation of powder metallurgy Ti-Nb-Ta-Zr-Al high entropy alloys. *Intermetallics*, 2018, 100, 95–103, http://doi.org/10.1016/j.intermet.2018.06.007.

[14] Prieto, E.; Oro Calderon, R. de; Konegger, T.; Gordo, E.; Gierl-Mayer, C.; Sheikh, S.; Guo, S.; Danninger, H.; Milenkovic, S.; Alvaredo, P. Processing of a new high entropy alloy: AlCrFeMoNiTi. *Powder Metallurgy*, 2018, 61, 258–265, http://doi.org/10.1080/00 325899.2018.1457862.

[15] Koundinya, N.T.B.N.; Sajith Babu, C.; Sivaprasad, K.; Susila, P.; Kishore Babu, N.; Baburao, J. Phase evolution and thermal analysis of nanocrystalline AlCrCuFeNiZn high entropy alloy produced by mechanical alloying. *Journal of Materials Engineering and Performance*, 2013, 22, 3077–3084, http://doi.org/10.1007/s11665-013-0580-5.

[16] Varalakshmi, S.; Kamaraj, M.; Murty, B.S. Synthesis and characterization of nano-crystalline AlFeTiCrZnCu high entropy solid solution by mechanical alloying. *Journal of Alloys and Compounds*, 2008, 460, 253–257, http://doi.org/10.1016/j.jallcom.2007.05.104.

[17] Waseem, O.A.; Ryu, H.J. Powder metallurgy processing of a WxTaTiVCr high-entropy alloy and its derivative alloys for fusion material applications. *Scientific Reports*, 2017, 7, 1926, http://doi.org/10.1038/s41598-017-02168-3.

[18] Waseem, O.A.; Lee, J.; Lee, H.M.; Ryu, H.J. The effect of Ti on the sintering and mechanical properties of refractory high-entropy alloy TixWTaVCr fabricated via spark plasma sintering for fusion plasma-facing materials. *Materials Chemistry and Physics*, 2018, 210, 87–94, http://doi.org/10.1016/j.matchemphys.2017.06.054.

[19] Cao, Y.; Liu, Y.; Liu, B.; Zhang, W. Precipitation behavior during hot deformation of powder metallurgy Ti-Nb-Ta-Zr-Al high entropy alloys. *Intermetallics*, 2018, 100, 95–103, http://doi.org/10.1016/j.intermet.2018.06.007.

[20] Zhang, Y.; Ai, Y.; Chen, W.; Ouyang, S. Preparation and microstructure and properties of AlCuFeMnTiV lightweight high entropy alloy. *Journal of Alloys and Compounds*, 2022, 900, 163352, http://doi.org/10.1016/j.jallcom.2021.163352.

[21] Song, R.; Wei, L.; Yang, C.; Wu, S. Phase formation and strengthening mechanisms in a dual-phase nanocrystalline CrMnFeVTi high-entropy alloy with ultrahigh hard-ness. *Journal of Alloys and Compounds*, 2018, 744, 552–560, http://doi.org/10.1016/j.jallcom.2018.02.029.

[22] Raza, A.; Kang, B.; Lee, J.; Ryu, H.J.; Hong, S.H. Transition in microstructural and mechanical behavior by reduction of sigma-forming element content in a novel high entropy alloy. *Materials & Design*, 2018, 145, 11–19, http://doi.org/10.1016/j.matdes.2018.02.036.

[23] Raza, A.; Ryu, H.J.; Hong, S.H. Strength enhancement and density reduction by the addi-tion of Al in CrFeMoV based high-entropy alloy fabricated through powder metallurgy. *Materials & Design*, 2018, 157, 97–104, http://doi.org/10.1016/j.matdes.2018.07.023.

[24] Zhang, Y.; Ai, Y.; Chen, W.; Ouyang, S. Preparation and microstructure and properties of AlCuFeMnTiV lightweight high entropy alloy. *Journal of Alloys and Compounds*, 2022, 900, 163352, http://doi.org/10.1016/j.jallcom.2021.163352.

[25] Kang, B.; Kong, T.; Ryu, H.J.; Hong, S.H. Superior mechanical properties and strength-ening mechanisms of lightweight AlxCrNbVMo refractory high-entropy alloys (x = 0, 0.5, 1.0) fabricated by the powder metallurgy process. *Journal of Materials Science & Technology*, 2021, 69, 32–41, http://doi.org/10.1016/j.jmst.2020.07.012.

[26] Cao, Y.; Liu, Y.; Liu, B.; Zhang, W. Precipitation behavior during hot deformation of powder metallurgy Ti-Nb-Ta-Zr-Al high entropy alloys. *Intermetallics*, 2018, 100, 95–103, http://doi.org/10.1016/j.intermet.2018.06.007.

[27] Gao, M.C.; Yeh, J.-W.; Liaw, P.K.; Zhang, Y. *High-Entropy Alloys*. 2016: Springer.

[28] Murty, B.S.; Yeh, J.-W.; Ranganathan, S. *High-Entropy Alloys*. 2014: Butterworth Heinemann. ISBN: 0128005262.

[29] Alaneme, K.K.; Bodunrin, M.O.; Oke, S.R. Processing, alloy composition and phase transition effect on the mechanical and corrosion properties of high entropy alloys: A review. *Journal of Materials Research and Technology*, 2016, 5(4), 384–393.

[30] Ostovari Moghaddam, A.; Shaburova, N.A.; Samodurova, M.N.; Abdollahzadeh, A.; Trofimov, E.A. Additive manufacturing of high entropy alloys: A practical review. *Journal of Materials Science & Technology*, 2021, 77, 131–162, http://doi.org/10.1016/j. jmst.2020.11.029.

[31] Kranz, J.; Herzog, D.; Emmelmann, C. Design guidelines for laser additive manufacturing of lightweight structures in TiAl6V4. *Journal of Laser Applications*, 2015, 27, S14001, http://doi.org/10.2351/1.4885235.

[32] Li, N.; Huang, S.; Zhang, G.; Qin, R.; Liu, W.; Xiong, H.; Shi, G.; Blackburn, J. Progress in additive manufacturing on new materials: A review. *Journal of Materials Science & Technology*, 2019, 35, 242–269, http://doi.org/10.1016/j.jmst.2018.09.002.

[33] Huang, H.; Wu, Y.; He, J.; Wang, H.; Liu, X.; An, K.; Wu, W.; Lu, Z. Phase-transformation ductilization of brittle high-entropy alloys via metastability engineering. *Advanced Materials*, 2017, 29, http://doi.org/10.1002/adma.201701678.

[34] Popov, V.V.; Katz-Demyanetz, A.; Koptyug, A.; Bamberger, M. Selective electron beam melting of Al0.5CrMoNbTa0.5 high entropy alloys using elemental powder blend. *Heliyon*, 2019, 5, e01188, http://doi.org/10.1016/j.heliyon.2019.e01188.

[35] Senkov, O.N.; Miracle, D.B.; Chaput, K.J.; Couzinie, J.-P. Development and exploration of refractory high entropy alloys—A review. *Journal of Materials Research*, 2018, 33, 3092–3128, http://doi.org/10.1557/jmr.2018.153.

[36] Zhang, H.; Zhao, Y.; Huang, S.; Zhu, S.; Wang, F.; Li, D. Manufacturing and analysis of high-performance refractory high-entropy alloy via selective laser melting (SLM). *Materials (Basel)*, 2019, 12, http://doi.org/10.3390/ma12050720.

[37] Dobbelstein, H.; Thiele, M.; Gurevich, E.L.; George, E.P.; Ostendorf, A. Direct metal deposition of refractory high entropy alloy MoNbTaW. *Physics Procedia*, 2016, 83, 624–633, http://doi.org/10.1016/j.phpro.2016.08.065.

[38] Dobbelstein, H.; Gurevich, E.L.; George, E.P.; Ostendorf, A.; Laplanche, G. Laser metal deposition of compositionally graded TiZrNbTa refractory high-entropy alloys using elemental powder blends. *Additive Manufacturing*, 2019, 25, 252–262, http://doi. org/10.1016/j.addma.2018.10.042.

[39] Zhou, R.; Liu, Y.; Liu, B.; Li, J.; Fang, Q. Precipitation behavior of selective laser melted FeCoCrNiC0.05 high entropy alloy. *Intermetallics*, 2019, 106, 20–25, http://doi. org/10.1016/j.intermet.2018.12.001.

[40] Xu, Z.; Zhang, H.; Li, W.; Mao, A.; Wang, L.; Song, G.; He, Y. Microstructure and nanoindentation creep behavior of CoCrFeMnNi high-entropy alloy fabricated by selective laser melting. *Additive Manufacturing*, 2019, 28, 766–771, http://doi.org/10.1016/j. addma.2019.06.012.

[41] Park, J.M.; Choe, J.; Kim, J.G.; Bae, J.W.; Moon, J.; Yang, S.; Kim, K.T.; Yu, J.-H.; Kim, H.S. Superior tensile properties of 1%C-CoCrFeMnNi high-entropy alloy additively manufactured by selective laser melting. *Materials Research Letters*, 2020, 8, 1–7, http://doi.org/10.1080/21663831.2019.1638844.

[42] Zhang, H.; Xu, W.; Xu, Y.; Lu, Z.; Li, D. The thermal-mechanical behavior of WTaMoNb high-entropy alloy via selective laser melting (SLM): Experiment and simulation. *The International Journal of Advanced Manufacturing Technology*, 2018, 96, 461–474, http://doi.org/10.1007/s00170-017-1331-9.

[43] Wang, P.; Huang, P.; Ng, F.L.; Sin, W.J.; Lu, S.; Nai, M.L.S.; Dong, Z.; Wei, J. Additively manufactured CoCrFeNiMn high-entropy alloy via pre-alloyed powder. *Materials & Design*, 2019, 168, 107576, http://doi.org/10.1016/j.matdes.2018.107576.

[44] Shiratori, H.; Fujieda, T.; Yamanaka, K.; Koizumi, Y.; Kuwabara, K.; Kato, T.; Chiba, A. Relationship between the microstructure and mechanical properties of an equiatomic AlCoCrFeNi high-entropy alloy fabricated by selective electron beam melting. *Materials Science and Engineering: A*, 2016, 656, 39–46, http://doi.org/10.1016/j.msea.2016.01.019.

[45] Kuwabara, K.; Shiratori, H.; Fujieda, T.; Yamanaka, K.; Koizumi, Y.; Chiba, A. Mechanical and corrosion properties of AlCoCrFeNi high-entropy alloy fabricated with selective electron beam melting. *Additive Manufacturing*, 2018, 23, 264–271, http://doi.org/10.1016/j.addma.2018.06.006.

[46] He, J.Y.; Liu, W.H.; Wang, H.; Wu, Y.; Liu, X.J.; Nieh, T.G.; Lu, Z.P. Effects of Al addition on structural evolution and tensile properties of the FeCoNiCrMn high-entropy alloy system. *Acta Materialia*, 2014, 62, 105–113, http://doi.org/10.1016/j.actamat.2013.09.037.

[47] Murty, B.S.; Yeh, J.-W.; Ranganathan, S. *High-Entropy Alloys*. 2014: Butterworth Heinemann. ISBN: 0128005262.

[48] Gao, M.C.; Yeh, J.-W.; Liaw, P.K.; Zhang, Y. *High-Entropy Alloys*. 2016: Springer.

[49] Rösler, J.; Harders, H.; Baeker, M. Mechanical behaviour of engineering materials: Metals, ceramics, polymers, and composites. *Springer Science & Business Media*, 2007, 108–109.

[50] Tsai, C.W.; Tsai, M.H.; Yeh, J.W.; Yang, C.C. Effect of temperature on mechanical properties of Al0.5CoCrCuFeNi wrought alloy. *Journal of Alloys and Compounds*, 2010, 490, 160–165.

[51] He, J.Y.; Liu, W.H.; Wang, H.; Wu, Y.; Liu, X.J.; Nieh, T.G.; Lu, Z.P. Effects of Al addition on structural evolution and tensile properties of the FeCoNiCrMn high-entropy alloy system. *Acta Materialia*, 2014, 62, 105–113.

[52] Senkov, O.N.; Wilks, G.B.; Miracle, D.B.; Chuang, C.P.; Liaw, P.K. Refractory high-entropy alloys. *Intermetallics*, 2010, 18, 1758–1765, http://doi.org/10.1016/j.intermet.2010.05.014.

[53] Senkov, O.N.; Scott, J.M.; Senkova, S.V.; Meisenkothen, F.; Miracle, D.B.; Woodward, C.F. Microstructure and elevated temperature properties of a refractory TaNbHfZrTi alloy. *Journal of Materials Science*, 2012, 47, 4062–4074, http://doi.org/10.1007/s10853-012-6260-2.

[54] Senkov, O.N.; Wilks, G.B.; Scott, J.M.; Miracle, D.B. Mechanical properties of Nb25Mo25Ta25W25 and V20Nb20Mo20Ta20W20 refractory high entropy alloys. *Intermetallics*, 2011, 19, 698–706, http://doi.org/10.1016/j.intermet.2011.01.004.

[55] Senkov, O.N.; Wilks, G.B.; Scott, J.M.; Miracle, D.B. Mechanical properties of Nb25Mo25Ta25W25 and V20Nb20Mo20Ta20W20 refractory high entropy alloys. *Intermetallics*, 2011, 19, 698–706.

[56] Yao, H.W.; Qiao, J.W.; Gao, M.C.; Hawk, J.A.; Ma, S.G.; Zhou, H.F.; Zhang, Y. NbTaV-(Ti,W) refractory high-entropy alloys: Experiments and modeling. *Materials Science and Engineering: A*, 2016, 674, 203–211, http://doi.org/10.1016/j.msea.2016.07.102.

[57] Wang, Y.P.; Li, B.S.; Ren, M.X.; Yang, C.; Fu, H.Z. Microstructure and compressive properties of AlCrFeCoNi high entropy alloy. *Materials Science and Engineering: A*, 2008, 491, 154–158, http://doi.org/10.1016/j.msea.2008.01.064.

[58] Wang, M.; Ma, Z.L.; Xu, Z.Q.; Cheng, X.W. Designing V NbMoTa refractory high-entropy alloys with improved properties for high-temperature applications. *Scripta Materialia*, 2021, 191, 131–136, http://doi.org/10.1016/j.scriptamat.2020.09.027.

[59] Senkov, O.N.; Semiatin, S.L. Microstructure and properties of a refractory high-entropy alloy after cold working. *Journal of Alloys and Compounds*, 2015, 649, 1110–1123.

[60] Lilensten, L.; Couzinie, J.P.; Perriere, L.; Hocini, A.; Keller, C.; Dirras, G.; Guillot, I. Study of a bcc multi-principal element alloy: Tensile and simple shear properties and underlying deformation mechanisms. *Acta Materialia*, 2018, 142, 131–141.

[61] Salem, A.A.; Kalidindi, S.R.; Doherty, R.D. Strain hardening of titanium: Role of deformation twinning. *Acta Mater*, 2003, 51, 4225–4237.

[62] Gutierrez-Urrutia, I.; Raabe, D. Dislocation and twin substructure evolution during strain hardening of an Fe–22wt.% Mn–0.6wt.% C TWIP steel observed by electron channeling contrast imaging. *Acta Mater*, 2011, 59, 6449–6462.

[63] Juan, C.-C.; Tsai, M.-H.; Tsai, C.-W.; Hsu, W.-L.; Lin, C.-M.; Chen, S.-K.; Lin, S.-J.; Yeh, J.-W. Simultaneously increasing the strength and ductility of a refractory high-entropy alloy via grain refining. *Materials Letters*, 2016, 184, 200–203.

[64] Tyson, W.R. Strengthening of hcp Zr, Ti and Hf by interstitial solutes—a review. *Canadian Metallurgical Quarterly*, 1967, 6, 301–332.

[65] Yu, Q.; Qi, L.; Tsuru, T.; Traylor, R.; Rugg, D.; Morris, J.W.; Asta, M.; Chrzan, D.C.; Minor, A.M. Origin of dramatic oxygen solute strengthening effect in titanium. *Science*, 2015, 347, 635–639.

[66] Gavriljuk, V.G. Influence of interstitial carbon, nitrogen, and hydrogen on the plasticity and brittleness of steel. *Steel in Translation*, 2015, 45, 747–753.

[67] Matsui, I.; Uesugi, T.; Takigawa, Y.; Higashi, K. Effect of interstitial carbon on the mechanical properties of electrodeposited bulk nanocrystalline Ni. *Acta Mater*, 2013, 61, 3360–3369.

[68] Shang, H.; Ma, B.; Shi, K.; Li, R.; Li, G. The strengthening effect of boron interstitial supersaturated solid solution on aluminum films. *Materials Letters*, 2017, 192, 104–106.

[69] Song, M.; Zhou, R.; Gu, J.; Wang, Z.; Ni, S.; Liu, Y. Nitrogen induced heterogeneous structures overcome strength-ductility trade-off in an additively manufactured high-entropy alloy. *Applied Materials Today*, 2020, 18, 100498.

[70] Baker, I. Interstitials in FCC high entropy alloys. *Metals*, 2020, 10, 695.

[71] Luo, H.; Li, Z.; Raabe, D. Hydrogen enhances strength and ductility of an equiatomic highentropy alloy. *Scientific Reports*, 2017, 7, 9892.

[72] Seol, J.B.; Bae, J.W.; Li, Z.; Chan Han, J.; Kim, J.G.; Raabe, D.; Kim, H.S. Boron doped ultrastrong and ductile high-entropy alloys. *Acta Materialia*, 2018, 151, 366–376.

[73] Wang, Z.; Baker, I.; Guo, W.; Poplawsky, J.D. The effect of carbon on the microstructures, mechanical properties, and deformation mechanisms of thermo-mechanically treated Fe40.4Ni11.3Mn34.8Al7.5Cr6 high entropy alloys. *Acta Materialia*, 2017, 126, 346–360.

[74] Wu, Z.; Parish, C.M.; Bei, H. Nano-twin mediated plasticity in carbon-containing FeNiCoCrMn high entropy alloys. *Journal of Alloys and Compounds*, 2015, 647, 815–822.

[75] Chen, J.; Yao, Z.; Wang, X.; Lu, Y.; Wang, X.; Liu, Y.; Fan, X. Effect of C content on microstructure and tensile properties of as-cast CoCrFeMnNi high entropy alloy. *Materials Chemistry and Physics*, 2018, 210, 136–145.

[76] Ye, Y.X.; Ouyang, B.; Liu, C.Z.; Duscher, G.J.; Nieh, T.G. Effect of interstitial oxygen and nitrogen on incipient plasticity of NbTiZrHf high-entropy alloys. *Acta Materialia*, 2020, 199, 413–424.

[77] Lei, Z.; Wu, Y.; He, J.; Liu, X.; Wang, H.; Jiang, S.; Gu, L.; Zhang, Q.; Gault, B.; Raabe, D.; Lu, Z. Snoek-type damping performance in strong and ductile high-entropy alloys. *Science Advances*, 2020, 6, 7802.

[78] Li, Z. Interstitial equiatomic CoCrFeMnNi high-entropy alloys: Carbon content, microstructure, and compositional homogeneity effects on deformation behavior. *Acta Materialia*, 2019, 164, 400–412.

[79] Casillas-Trujillo, L.; Jansson, U.; Sahlberg, M.; Ek, G.; Nygård, M.M.; Sørby, M.H.; Hauback, B.C.; Abrikosov, I.; Alling, B. Interstitial carbon in bcc HfNbTiVZr high entropy alloy from first principles. *arXiv preprint arXiv*, 2020, 2010, 01354.

[80] Moravcik, I.; Hadraba, H.; Li, L.; Dlouhy, I.; Raabe, D.; Li, Z. Yield strength increase of a CoCrNi medium entropy alloy by interstitial nitrogen doping at maintained ductility. *Scripta Materialia*, 2020, 178, 391–397.

[81] Seol, J.B.; Bae, J.W.; Kim, J.G.; Sung, H.; Li, Z.; Lee, H.H.; Shim, S.H.; Jang, J.H.; Ko, W.-S.; Hong, S.I.; Kim, H.S. Short-range order strengthening in boron-doped high-entropy alloys for cryogenic applications. *Acta Materialia*, 2020, 194, 366–377.

[82] Chen, Y.; Li, Y.; Cheng, X.; Xu, Z.; Wu, C.; Cheng, B.; Wang, M. Interstitial strengthening of refractory ZrTiHfNb0.5Ta0.5Ox (x= 0.05, 0.1, 0.2) high-entropy alloys. *Materials Letters*, 2018, 228, 145–147.

[83] Lei, Z.F.; Liu, X.J.; Wu, Y.; Wang, H.; Jiang, S.H.; Wang, S.D.; Hui, X.D.; Wu, Y.D.; Gault, B.; Kontis, P.; Raabe, D.; Gu, L.; Zhang, Q.H.; Chen, H.W.; Wang, H.T.; Liu, J.B.; An, K.; Zeng, Q.S.; Nieh, T.G.; Lu, Z.P. Enhanced strength and ductility in a high-entropy alloy via ordered oxygen complexes. *Nature*, 2018, 563, 546.

[84] Huang, H.; Wu, Y.; He, J.; Wang, H.; Liu, X.; An, K.; Wu, W.; Lu, Z. Phase-transformation ductilization of brittle high-entropy alloys via metastability engineering. *Advanced Materials*, 2017, 29, 1701678.

[85] Han, Z.D.; Chen, N.; Zhao, S.F.; Fan, L.W.; Yang, G.N.; Shao, Y.; Yao, K.F. Effect of Ti additions on mechanical properties of NbMoTaW and VNbMoTaW refractory high entropy alloys. *Intermetallics*, 2017, 84, 153–157, http://doi.org/10.1016/j.intermet.2017.01.007.

[86] Senkov, O.; Scott, J.; Senkova, S.; Meisenkothen, F.; Miracle, D.; Woodward, C. Microstructure and elevated temperature properties of a refractory TaNbHfZrTi alloy. *Journal of Materials Science*, 2012, 47, 4062–4074.

[87] Hanamura, T.; Yin, F.; Nagai, K. Ductile-brittle transition temperature of ultrafine ferrite/cementite microstructure in a low carbon steel controlled by effective grain size. *ISIJ International*, 2004, 44, 610–617.

[88] Trinkle, D.R.; Woodward, C. The chemistry of deformation: How solutes soften pure metals. *Science*, 2005, 310, 1665.

[89] Zhang, L.; Chen, X.; Huang, Y.; Liu, W.; Ma, Y. Microstructural characteristics and evolution mechanisms of 90W—Ni—Fe alloy under high-strain-rate deformation. *Materials Science and Engineering: A*, 2021, 811, 141070, http://doi.org/10.1016/j.msea.2021.141070.

[90] Luo, R.; Huang, D.; Yang, M.; Tang, E.; Wang, M.; He, L. Penetrating performance and "self-sharpening" behavior of fi-ne-grained tungsten heavy alloy rod penetrators. *Materials Science and Engineering: A*, 2016, 675, 262–270, http://doi.org/10.1016/j.msea.2016.08.060.

[91] Liu, X.-F.; Tian, Z.-L.; Zhang, X.-F.; Chen, H.-H.; Liu, T.-W.; Chen, Y.; Wang, Y.-J.; Dai, L.-H. "Self-sharpening" tungsten high-entropy alloy. *Acta Materialia*, 2020, 186, 257–266, http://doi.org/10.1016/j.actamat.2020.01.005.

[92] Zhang, L.; Chen, X.; Huang, Y.; Liu, W.; Ma, Y. Microstructural characteristics and evolution mechanisms of 90W-Ni-Fe alloy under high-strain-rate deformation. *Materials Science and Engineering: A*, 2021, 811, 141070, http://doi.org/10.1016/j.msea.2021.141070.

[93] Müller, F.; Gorr, B.; Christ, H.-J.; Müller, J.; Butz, B.; Chen, H.; Kauffmann, A.; Heilmaier, M. On the oxidation mechanism of refractory high entropy alloys. *Corrosion Science*, 2019, 159, 108161, http://doi.org/10.1016/j.corsci.2019.108161.

[94] Qiu, Y.; Thomas, S.; Gibson, M.A.; Fraser, H.L.; Birbilis, N. Corrosion of high entropy alloys. *NPJ Mater Degrad*, 2017, 1, http://doi.org/10.1038/s41529-017-0009-y.

[95] Jensen, J.K.; Welk, B.A.; Williams, R.; Sosa, J.M.; Huber, D.E.; Senkov, O.N.; Viswanathan, G.B.; Fraser, H.L. Characterization of the microstructure of the compositionally complex alloy Al1Mo0.5Nb1Ta0.5Ti1Zr1. *Scripta Materialia*, 2016, 121, 1–4, http://doi.org/10.1016/j.scriptamat.2016.04.017.

[96] Fu, Y.; Li, J.; Luo, H.; Du, C.; Li, X. Recent advances on environmental corrosion behavior and mechanism of high-entropy alloys. *Journal of Materials Science & Technology*, 2021, 80, 217–233, http://doi.org/10.1016/j.jmst.2020.11.044.

[97] Wang, S.-P.; Xu, J. TiZrNbTaMo high-entropy alloy designed for orthopedic implants: As-cast microstructure and mechanical properties. *Materials Science & Engineering C-Materials for Biological Applications*, 2017, 73, 80–89, http://doi.org/10.1016/j.msec.2016.12.057.

[98] El-Atwani, O.; Li, N.; Li, M.; Devaraj, A.; Baldwin, J.K.S.; Schneider, M.M.; Martinez, E. Outstanding radiation re-sistance of tungsten-based high-entropy alloys. *Science Advances*, 2019, 5(3), eaav2002, http://doi.org/10.1126/sciadv.aav2002.

[99] Zhang, Z.; Han, E.-H.; Xiang, C. Irradiation behaviors of two novel single-phase bcc-structure high-entropy alloys for accident-tolerant fuel cladding. *Journal of Materials Science & Technology*, 2021, 84, 230–238, http://doi.org/10.1016/j.jmst.2020.12.058.

4 Multiple-Phase High-Entropy Materials

Yong Zhang and Shichao Zhou

4.1 INTRODUCTION

High-entropy alloys (HEAs) with a single phase usually possess alternative mechanical performance (i.e., high strength or good ductility). Such a conflict between strength and ductility is viewed as a trade-off [1]. To overcome this mutual exclusion, many researchers have paid a great endeavor to relieve such a grim case [2]. Among them, architecting the microstructure from single phase to duplex phase achieved an effective enhancement of the synergism between the strength and ductility (i.e., overcoming the strength-ductility trade-off) [3].

4.2 MICROSTRUCTURES AND PROPERTIES OF DUPLEX-PHASE HEAs

Ma et al. [4] investigated the microstructures and performances of the as-cast AlCoCrFeNiNb$_x$ (x = 0, 0.1, 0.25, 0.5, and 0.75) high-entropy alloys. As shown in Figure 4.1, AlCoCrFeNiNb$_{0.1}$ (x = 0.1) possessed typical body-centered cubic (BCC) dendritic with rich Al, Ni, and Laves inter-dendritic region with rich Cr, Fe, Nb (Laves can be identified as a hexagonal close-packed, HCP, lattice structure). With the Nb increase, AlCoCrFeNiNb$_{0.25}$ and AlCoCrFeNiNb$_{0.5}$, the dendrite and inter-dendrite transited primary BCC phase and eutectic phases with alternative lamellar BCC and Laves phases. This phenomenon is due to the Nb not just exhibiting very negative enthalpies of mix with other incorporated elements (the values of Nb and Al, Co, Cr, Fe, and Ni atomic pairs are −18, −25, −7, −16, and −30 kJ/mol, separately) but also possessing the largest atomic size. Both of which accelerated the formation of the secondary phase. With the further increase of Nb, AlCoCrFeNiNb$_{0.75}$ showed primary Laves phase and eutectic phases with alternative lamellar BCC and Laves phases. Such a transition can be viewed as hypoeutectic to hypereutectic.

The following corresponds to the phase formation rule for HEAs defined by Zhang [5]:

Industrial materials usually contain many phases to keep their comprehensive performance properties. Casting alloys usually are eutectic composition alloys; the ductile primary phase may also make the alloy low cost and have higher properties.

DOI: 10.1201/9781003319986-4

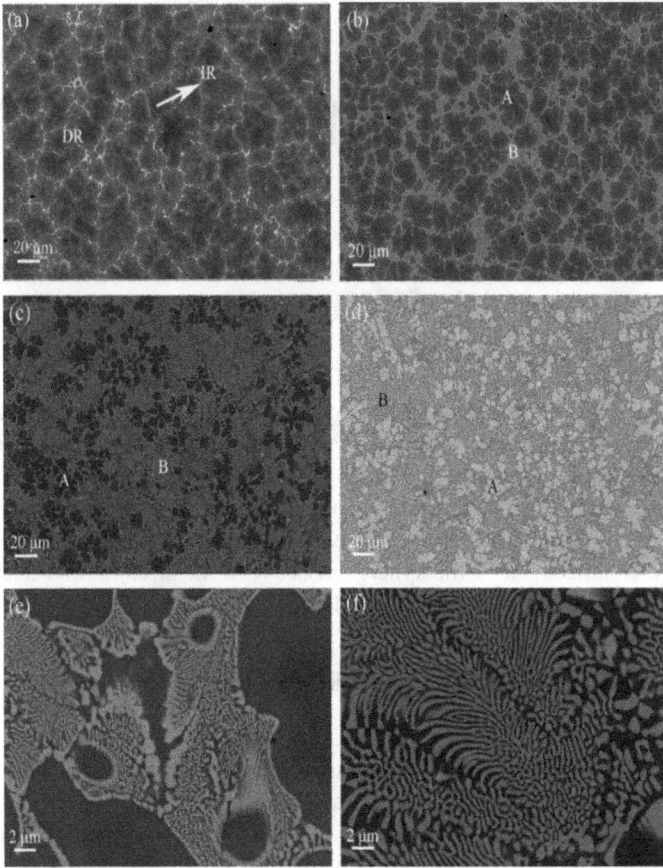

FIGURE 4.1 Backscatter electron images of the AlCoCrFeNiNb$_x$ alloys, x = 0.1, 0.25, 0.5, and 0.75, corresponding to (a), (b), (c), and (d), respectively; (e) and (f) are the magnification of B region in (b) and (c), respectively.

$$\delta = \sqrt{\sum_{i=1}^{n} c_i \left(1 - \frac{r_i}{\bar{r}}\right)^2} \tag{1}$$

$$\Omega = T_m \Delta S_{mix} / \left| \Delta H_{mix} \right| \tag{2}$$

$$\Delta H_{mix} = \sum_{i=1,i \neq j}^{n} c_i c_j \Omega_{ij} \tag{3}$$

$$\Delta S_{mix} = -R \sum_{i=1}^{n} c_i \ln c_i \tag{4}$$

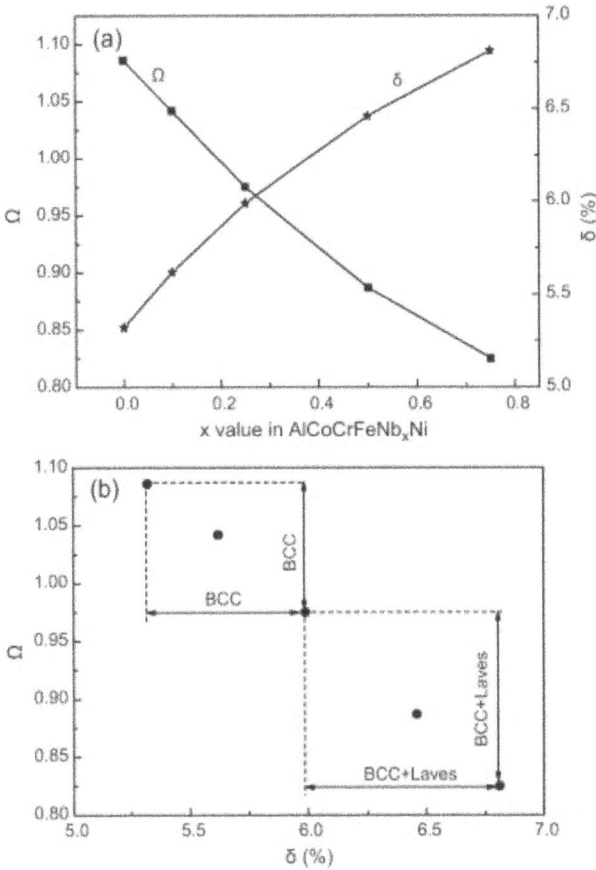

FIGURE 4.2 (a) The curves of Ω and δ as a function of Nb contents for the AlCoCrFeNiNb$_x$ alloys ($x = 0, 0.1, 0.25, 0.5,$ and 0.75); (b) the correlation between Ω and δ.

Where $\bar{r} = \sum_{i=1}^{n} c_i r_i$, c_i and r_i are the atomic percentage of the ith component; in $T_m = \sum_{i=1}^{n} c_i \left(T_m \right)_i$, $(T_m)_i$ is the atomic melting temperature of the ith component; in $\Omega_{ij} = 4\Delta H_{mix}^{AB}$, ΔH_{mix}^{AB} is the regular solution interaction parameter between the ith and jth elements; the curves of Ω and δ as a function of Nb contents and the correlation between Ω and δ are shown in Figure 4.2 The increase of Nb in this alloy system led to larger atomic size differences, while the decreasing Ω indicated that the entropy of mixing played a negative role compared to the enthalpy of mixing in this alloy system.

By designing this duplex-phase structure in the AlCoCrFeNiNb$_x$ ($x = 0, 0.1, 0.25,$ 0.5, and 0.75), an enhanced synergism of the yield strength and plastic strain under compress tests is displayed in the AlCoCrFeNiNb$_{0.1}$ and AlCoCrFeNiNb$_{0.25}$, as shown in Figure 4.3.

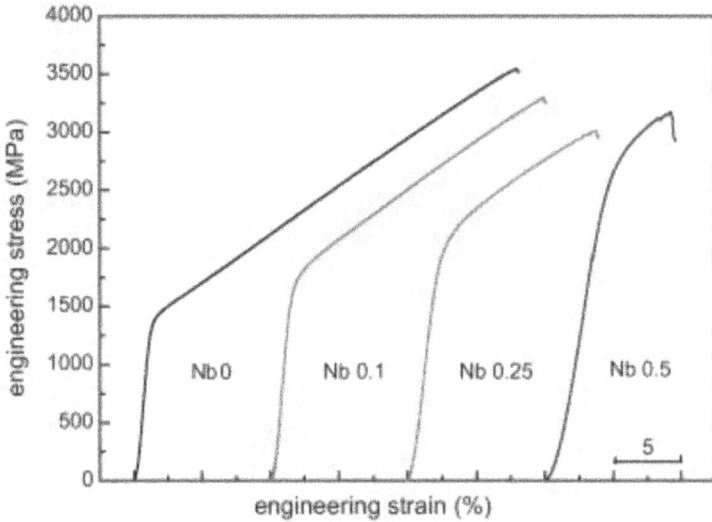

FIGURE 4.3 The compressive stress-strain curves of the AlCoCrFeNiNb$_x$ (x = 0, 0.1, 0.25, and 0.5) HEAs.

Similar to the earlier study, Zhang et al. [6] constructed a good combination of strength and ductility in the FeCoNi(AlSi)$_x$ (x = 0, 0.1, 0.2, 0.3, 0.4, 0.5, and 0.8) HEAs. The increase of Al and Si in this system facilitated the formation of a BCC phase with enriched Ni, Si, and Al atoms from the basic face-centered cubic (FCC) phases with enriched Fe and Co in the as-cast state. However, as shown in Figure 4.4, the low abundance of Al and Si led to a bad yield strength, high ultimate strength, and excellent ductility under compress tests. Opposite to that, exorbitant Al and Si caused the increase of yield strength yet severe plasticity decline due to a massive BCC phase hastened local strain and necking. The direct evidence is displayed in Figure 4.5.

Figure 4.5 shows the microstructures of the as-cast FeCoNi(AlSi)$_x$ (x = 0, 0.1, 0.2, 0.3, 0.4, 0.5, and 0.8) HEAs. With the increase of Al and Si (x = 0, 0.1, 0.2, 0.3), the single FCC phase transits to FCC and BCC phase (corresponding dendrite and inter-dendrite). As x adds up to 0.8, the microstructure comprised a primary BCC phase and a slight embedded FCC phase. The increasing brittle BCC phase induced the decrease in ductility. Among them, FeCoNi(AlSi)$_{0.3}$ possessed the moderate yield strength, highest ultimate stress, and excellent plasticity (i.e., breaking through the strength-ductility trade-off).

He et al. [3] investigated systematically a series of the as-cast (FeCoNiCrMn)$_{100-}$$_xAl_x$ (x = 0–20 at.%) HEAs to explore the effect of Al on the microstructures and properties (Figure 4.6). As the Al concentration increased, the microstructure transited from the initial single FCC structure to a duplex FCC and BCC structure and then a single BCC structure. Resulting from the crystalline structural changes, there were corresponding variations in tensile properties. The single FCC (FeCoNiCrMn)$_{100-x}$Al$_x$ (x < 8%) alloys behaved like a solid solution with relatively low strength yet excellent

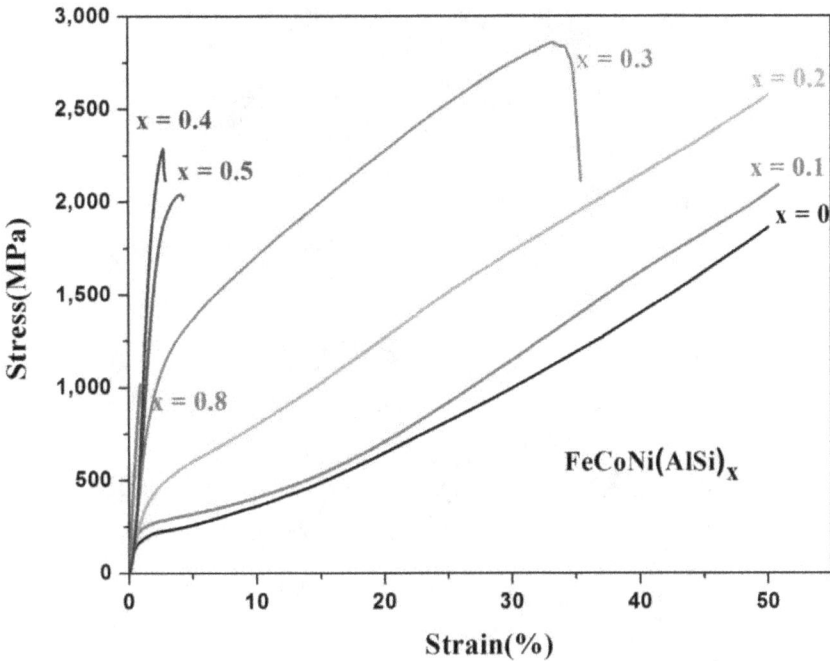

FIGURE 4.4 Engineering stress-strain curves of FeCoNi(AlSi)$_x$ EHEAs under compress test at room temperature.

ductility (tensile strength of ~500 MPa, yield strength of ~220 MPa, and elongation between 47.2% and 61.7%). The duplex FCC and BCC (FeCoNiCrMn)$_{100-x}$Al$_x$ ($x = 8\%$–16%) alloys achieved a composite with an increase in strength but reduced ductility (for instance, (FeCoNiCrMn)$_{89}$Al$_{11}$ possessed the highest tensile strength of 1174 MPa and 7.7% ductility). The single BCC (FeCoNiCrMn)$_{100-x}$Al$_x$ ($x > 16\%$) alloys became extremely brittle. Similar to the variation trend of the strength, the Vickers hardness also increased with the Al concentration raised.

Different from consistent stable dual-phase HEAs as introduced earlier, there is a series of Fe$_{(80-x)}$Mn$_x$Co$_{10}$Cr$_{10}$ (at%) HEAs with metastable structure [7]. The as-cast ingots were first hot-rolled at 900°C subsequently homogenized at 1200°C for 2 h followed by water-quenching. For the Fe$_{50}$Mn$_{30}$Co$_{10}$Cr$_{10}$ (at%) HEA, further grain refinement was realized by cold-rolling and annealing. With the increase of the Mn content, Fe$_{(80-x)}$Mn$_x$Co$_{10}$Cr$_{10}$ HEAs possessed decreased phase stability and gradually activated twinning-induced plasticity (TWIP) and transformation-induced plasticity (TRIP) effects. Furthermore, dislocation slip, twinning, and the formation of stacking faults induced the mechanical properties.

The mechanical behaviors of Fe$_{(80-x)}$Mn$_x$Co$_{10}$Cr$_{10}$ HEAs are very good. Compared with other single-phase HEAs and coarse-grain Fe$_{50}$Mn$_{30}$Co$_{10}$Cr$_{10}$, the fine-grained Fe$_{50}$Mn$_{30}$Co$_{10}$Cr$_{10}$ achieved an exceptional combination of strength and ductility, overcoming the strength-ductility trade-off.

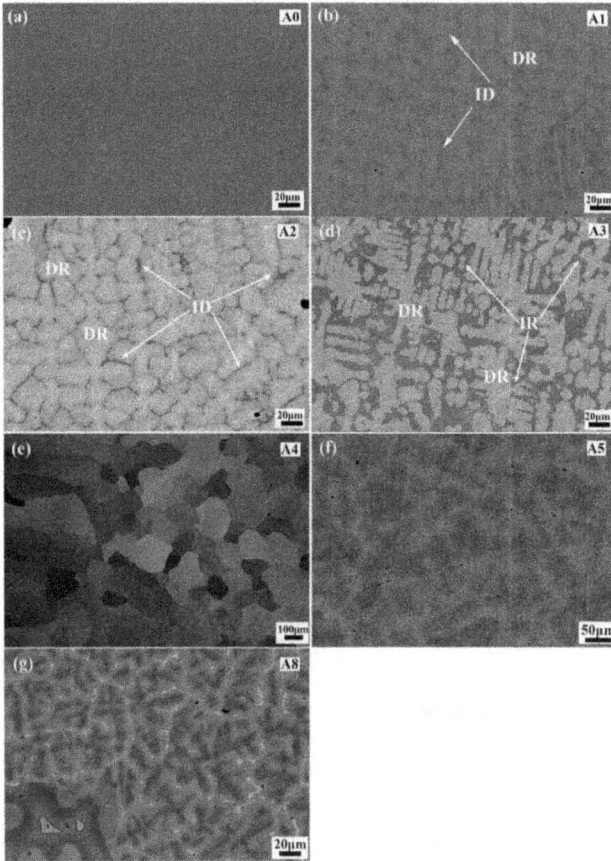

FIGURE 4.5 Microstructures of FeCoNi(AlSi)$_x$ ($x = 0$, 0.1, 0.2, 0.3, 0.4, 0.5, and 0.8) HEAs. SEM backscattering electron images of FeCoNi(AlSi)x alloys, (a) x=5, (b) x=0.1, (c) x=0.2, (d) x=0.3, (e) x=0.4, (f) x=0.5, and (g) x=0.8.

Inspired by this idea, Nene et al. [8] designed a duplex-phase Fe$_{42}$Mn$_{28}$Co$_{10}$Cr$_{15}$Si$_5$ by utilizing metastable engineering to introduce transformation and twinning during deformation. The addition of Si played an important role in this HEA; it not only decreased the stability of FCC but also enhanced the solid solubility, owing to the significant differences in atomic sizes. Such a decrease in the structure stability further promotes transformation-induced plastic deformation. Both the as-cast and improved state Fe$_{42}$Mn$_{28}$Co$_{10}$Cr$_{15}$Si$_5$ HEAs possessed similar compositions and phase structures, as displayed in Figure 4.7. The biggest difference between those two HEAs is the grain size. The grain size of the HEA with an improved state is ~1.3 μm, which is significantly finer than that in as-cast HEA (~100 μm).

This HEA achieved a high strength (~1.15 GPa) and ductility (~11%), as displayed in Figure 4.8, owing to the transformation-induced plasticity effect. As shown in Figure 4.8b, both the as-cast and improved state HEAs appeared to have an apparent transformation after deformation, which induced the strain-hardening.

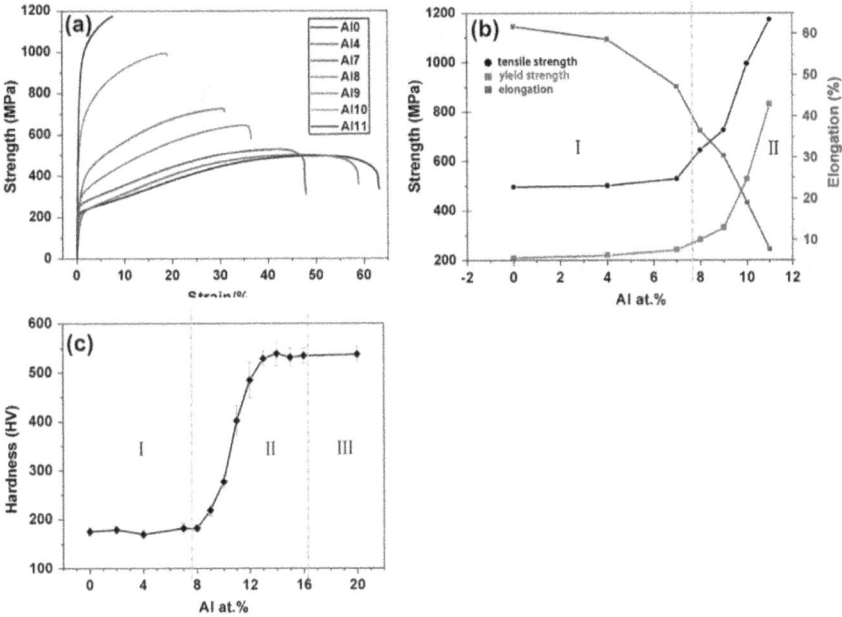

FIGURE 4.6 Mechanical properties of the as-cast Al$_x$ alloys: (a) engineering stress-strain curves; (b) tensile strength, yield strength, and elongation; (c) Vickers hardness variation with the Al content increase of (FeCoNiCrMn)$_{100-x}$Al$_x$.

Different from the conventional cast method, Li et al. [9] engineered a series of Al$_x$CoCrFeNi ($x = 0, 0.3, 0.5, 0.75$, and 1) HEAs through the high-gravity combustion from oxides. This route has been proven to produce kinds of ceramic materials, and it is the first application in HEAs.

It can be clearly seen that CoCrFeNi ($x = 0$) HEA exhibited the uniform distribution microstructure, as displayed in Figure 4.9. However, the element segregation and a typical dendritic and inter-dendritic morphology formed with the addition of the Al element. This trend can be explained by the significant differences between the Al and other elements. This CoCrFeNi alloy displayed a homogeneous structure as well as elemental distributions, exhibiting no influence by the high gravity, as shown in Figure 4.9(a). The Al$_{0.3}$CoCrFeNi ($x = 0.3$) HEA transited to the long-strip-shaped structure from the regular oval-shaped dendrites along the high gravity direction, as shown in Figure 4.9(b). Furthermore, coarse strip-shaped dendrites appeared in the Al$_{0.5}$CoCrFeNi ($x = 0.5$) HEA, and the homogeneous microstructures still formed, ignoring the effect of the high gravity (Figure 4.9(c)). The Al$_{0.75}$CoCrFeNi ($x = 0.75$) HEA exhibited homogeneous dendrites along the high gravity direction. For the AlCoCrFeNi alloy, as shown in Figure 4.9(e), the dendrite appeared a bit off course along the high gravity direction.

With the increase of the Al (Figure 4.10), the strength tended to increase while decreasing in plasticity for Al$_x$CoCrFeNi HEAs; that is, due to the crystal structure

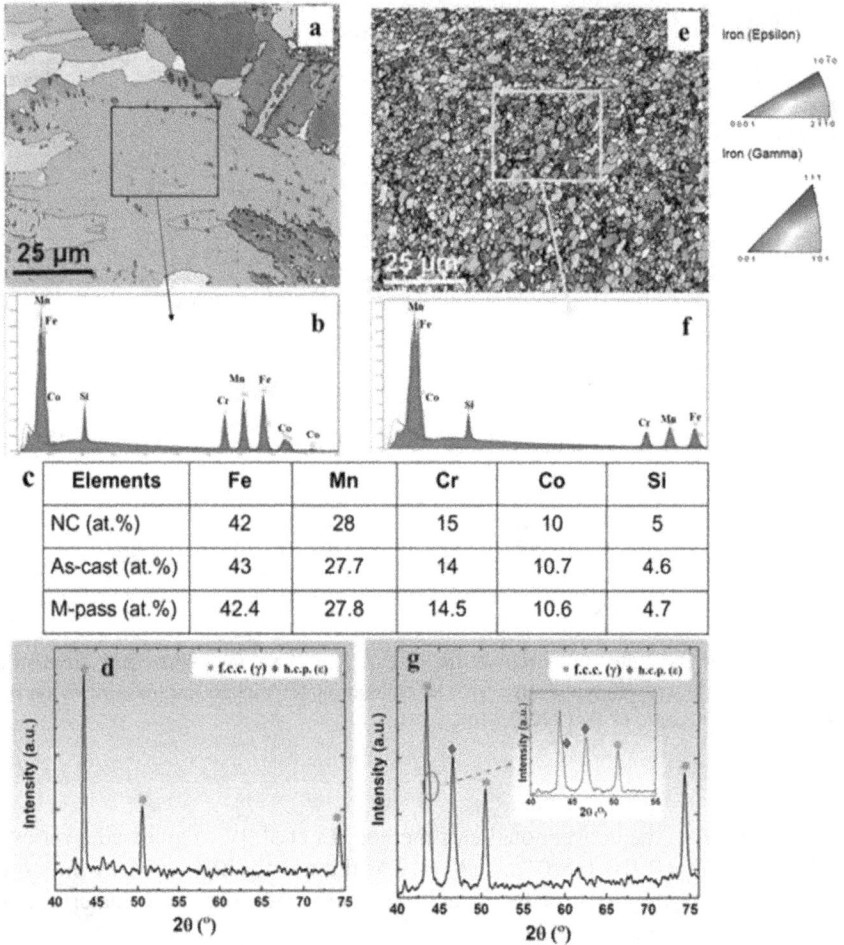

FIGURE 4.7 Microstructures of $Fe_{42}Mn_{28}Co_{10}Cr_{15}Si_5$ HEAs. (a–d) EBSD map, EDS result with elemental composition, and XRD for as-cast $Fe_{42}Mn_{28}Co_{10}Cr_{15}Si_5$ HEA; (e–g) EBSD map, EDS result with elemental composition, and XRD for improved-state $Fe_{42}Mn_{28}Co_{10}Cr_{15}Si_5$ HEA.

transition from FCC to BCC. The CoCrFeNi HEA possessed the lowest strength with the highest plastic strain. Both the $Al_{0.5}CoCrFeNi$ and $Al_{0.75}CoCrFeNi$ HEAs achieved an excellent combination of the ultimate fracture strength (1750 MPa and 1870 MPa) and plastic strains (42% and 22%), respectively, which resulted from the synergism of the hard BCC and soft FCC phases. With the further increase in the Al content, AlCoCrFeNi exhibited the lowest plastic strain and moderate strength owing to the decrease of strain-hardening.

Furthermore, Xia et al. [10] investigated the impact toughness of the as-cast $Al_xCoCrFeNi$ (x = 0, 0.1, 0.75, and 1.5) HEAs at 77 K, 200 K, and 298 K. The

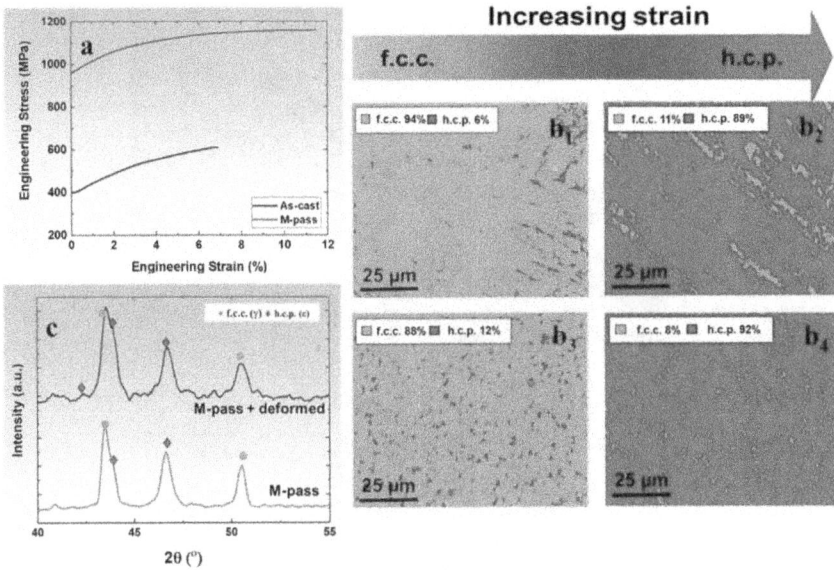

FIGURE 4.8 (a) Engineering stress-strain curves of $Fe_{42}Mn_{28}Co_{10}Cr_{15}Si_5$ HEAs at ambient temperature; (b1, b2) EBSD phase maps before and after tensile deformation of as-cast state; (b3, b4) EBSD phase maps before and after tensile deformation of optimized state; (c) XRD pattern.

AlCoCrFeNi and $Al_{0.1}$CoCrFeNi HEAs possessed an inverse temperature dependence behavior; that is, a much higher impact toughness at 77 K than 298 K. Such an abnormal phenomenon resulted from the formation of nano-twinning during the plastic deformation. However, the $Al_{0.75}$CoCrFeNi HEA achieved a moderate impact toughness compared with other HEAs.

4.3 IRRADIATION BEHAVIOR OF DUPLEX-PHASE HEAs

Xia et al. [11] studied the irradiation behavior of the as-cast Al_xCoCrFeNi ($x = 0.1$, 0.75, and 1.5) HEAs under a 3 MeV Au-ions irradiation. The result showed that $Al_{0.75}$CoCrFeNi HEA possessed a duplex-phase (FCC+BCC) structure, while $Al_{0.1}$CoCrFeNi and $Al_{1.5}$CoCrFeNi HEAs exhibited single FCC and BCC structure, respectively, corresponding to previous studies, as shown earlier. As shown in Figure 4.11, all of them possessed good structure stability and resistance to volume swelling induced by irradiation. The duplex-phase $Al_{0.75}$CoCrFeNi and single-phase FCC $Al_{0.1}$CoCrFeNi HEAs possessed an exceptional resistance to swelling, with a tiny height difference of 5 nm and 7.6 nm, separately.

Furthermore, Xia et al. [12] demonstrated there is no ordering trend, amorphization, and phase separation formed under heavy ion irradiation at room temperature to high dPa for these three HEAs.

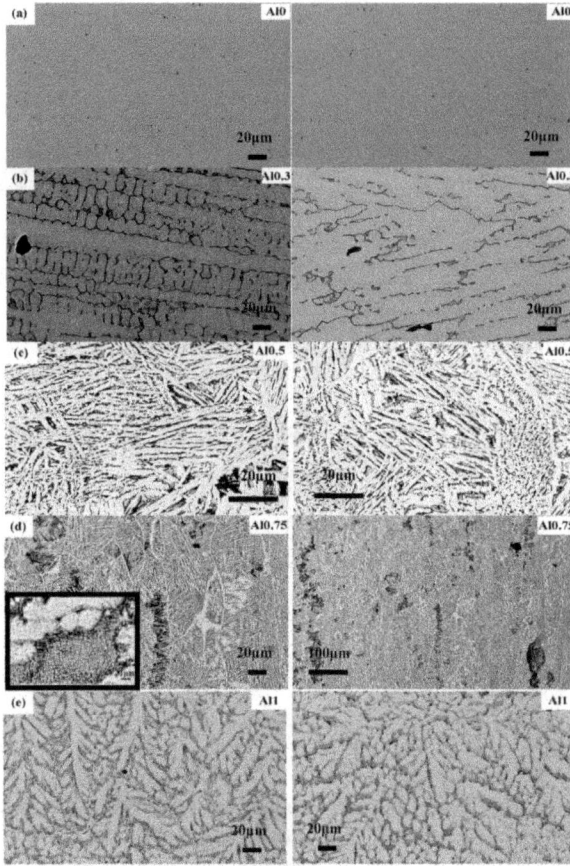

FIGURE 4.9 The SEM images of (a) CoCrFeNi, (b) $Al_{0.3}CoCrFeNi$, (c) $Al_{0.5}CoCrFeNi$, (d) $Al_{0.75}CoCrFeNi$, and (e) AlCoCrFeNi along the ultrahigh gravity direction.

4.4 EUTECTIC HEAs

Generally, the dual-phase HEAs exhibited dendritic and inter-dendritic morphology or disperse-phase distributions, as mentioned earlier. Interestingly, a series of eutectic HEAs (EHEAs) usually possessed a regular lamellar microstructure with dual phases at the as-cast state.

Lu et al. [13] innovatively developed the first EHEA, $AlCoCrFeNi_{2.1}$, which possesses excellent liquidity and castability as well as scarce casting flaws. As displayed in XRD pattern, FCC and B2 phases were identified, as shown in Figure 4.12(a). Both the heating and cooling curves of the differential scanning calorimetric (DSC) show only one melting peak, indicating the eutectic composition of this alloy (Figure 4.12(b)).

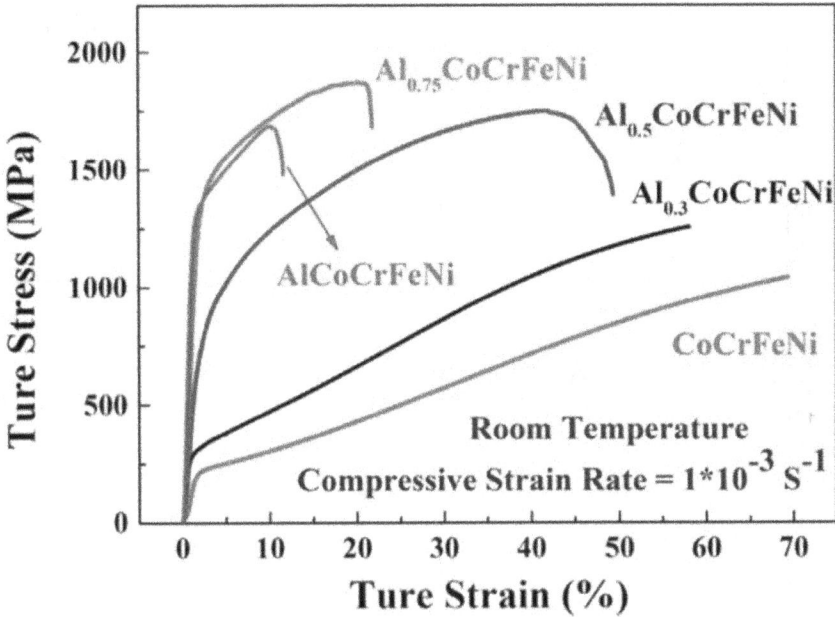

FIGURE 4.10 Compressive true stress-strain curves of Al$_x$CoCrFeNi (x = 0, 0.3, 0.5, 0.75, and 1) HEAs.

The typical lamellar structure is displayed in Figure 4.13(a). The enlarged image (Figure 4.13(b)) shows that the apparent dark/tint region in the micrometer scale are FCC (marked A) and B2 (marked B) phases, respectively. Moreover, there are B2 precipitates embedded in the FCC lamellar.

Figure 4.14 shows the stress-strain curves of the as-cast AlCoCrFeNi$_{2.1}$ EHEAs at room and elevated temperatures. The high ultimate strength (944 MPa) and ductility (25.6%) implied an excellent mechanical property at ambient temperature. Furthermore, at elevated temperatures of 600°C and 700°C, the ultimate stress and elongation were 806 MPa, 33.7%, and 538 MPa, 22.9%, respectively, which exhibited that an excellent synergy of the strength and ductility can be maintained up to 700°C.

However, the low yield strength (~100 MPa) of the AlCoCrFeNi$_{2.1}$ EHEA may limit its further application. Intriguingly, the high strain-hardening ability of this EHEA implies a great potential to enhance yield strength through the proper thermodynamic process.

Wani et al. [14] prepared an EHEA with an equiaxed duplex microstructure by 90% cold-rolled and annealed at 800°C for 1 hour, as displayed in Figure 4.15(a) and Figure 4.16. Such a microstructure is totally different from the alternative lamellar morphology at the as-cast state mentioned earlier. The TEM bright field image also shows an equiaxed grain with ultrafine-grained morphology (Figure 4.15(b)), and the corresponding SADPs are displayed in Figure 4.15(c).

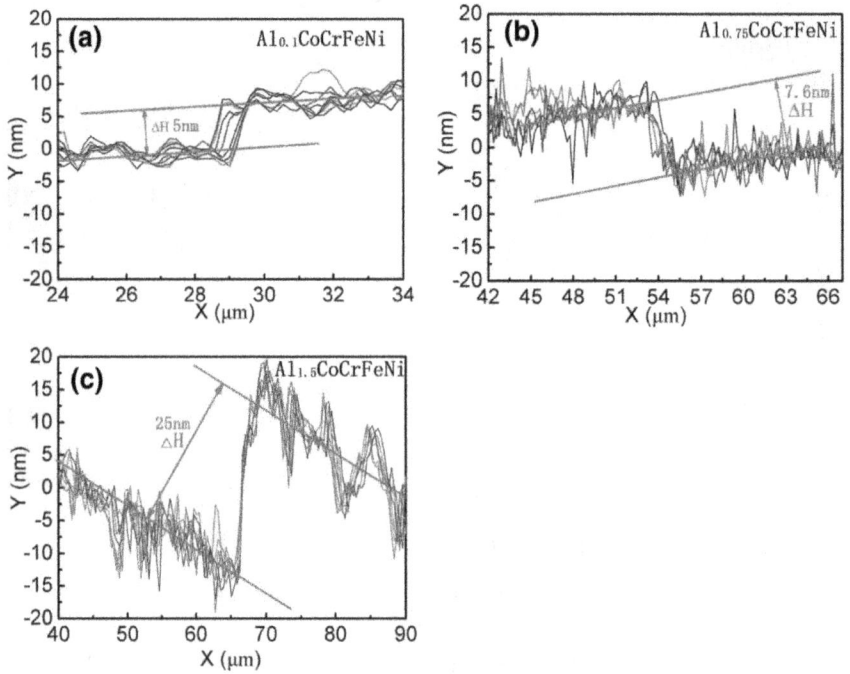

FIGURE 4.11 The height profile of lines across the boundary of irradiated and masked regions: (a) $Al_{0.1}CoCrFeNi$, (b) $Al_{0.75}CoCrFeNi$ (duplex phases), and (c) $Al_{1.5}CoCrFeNi$.

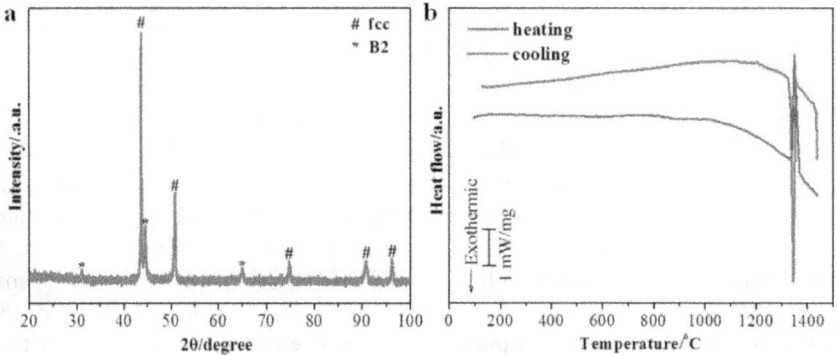

FIGURE 4.12 The XRD pattern (a) and DSC curve (b) of the as-cast $AlCoCrFeNi_{2.1}$ EHEA.

In order to further reinforce the mechanical properties, Shi et al. [15] architected ultrafine-grained $AlCoCrFeNi_{2.1}$ EHEA by deliberately inheriting the lamellae structure. As shown in Figure 4.17(b) and Figure 4.17(c), the customized EHEA displayed a characteristic lamella morphology, much like the as-cast EHEA, suggesting the

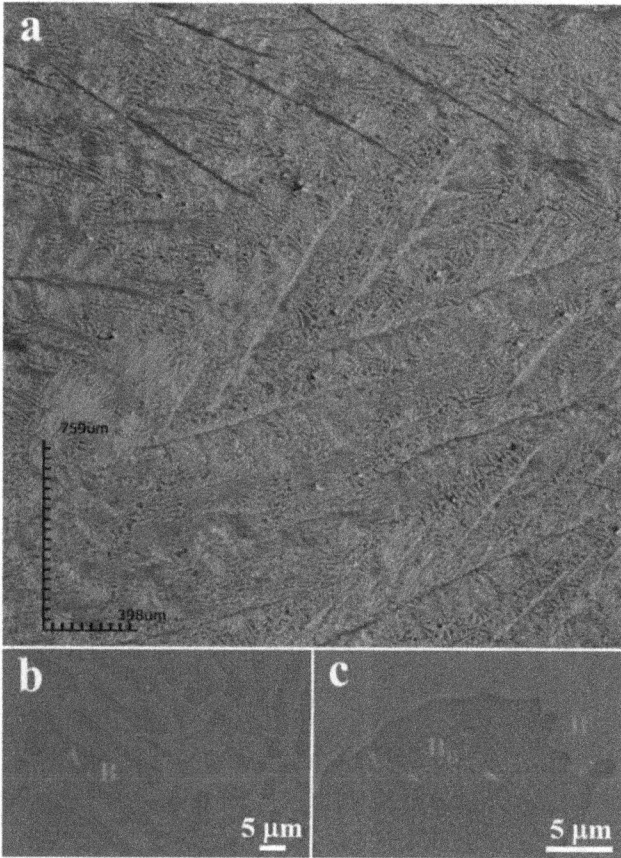

FIGURE 4.13 Microstructures of AlCoCrFeNi$_{2.1}$ EHEA: (a) the image showing the eutectic structure, (b) the enlarged SEM image showing the lamellar structure, and (c) the B2 precipitates in the FCC lamellar.

inheritance of a lamellar structure. Both the FCC and B2 phases comprised equiaxed grains (~0.71 μm). Moreover, this FCC lamellae contain precipitated B2 phases of various sizes, as displayed in Figure 4.17(e). This microstructure ingeniously combines ultrafine-grain and lamellae structures, which is different from an as-cast and equiaxed duplex structure.

The mechanical responses of this customized EHEA are displayed in Figure 4.18. Compared to other HEAs, such customized EHEAs (named DPHL 660, 700, and 740, respectively) possess significantly improved performances (Figure 4.18(a)); for instance, the yield strength, tensile strength, and ductility of DPHL 660 are 1,490 MPa, 1,638 MPa, and 16%. In addition, the distinguished strain-hardening is shown in Figure 4.18(b). The inset in Figure 4.18(a) shows the loading-unloading-reloading outcome, from which the apparent hysteresis loops implied the existence of a

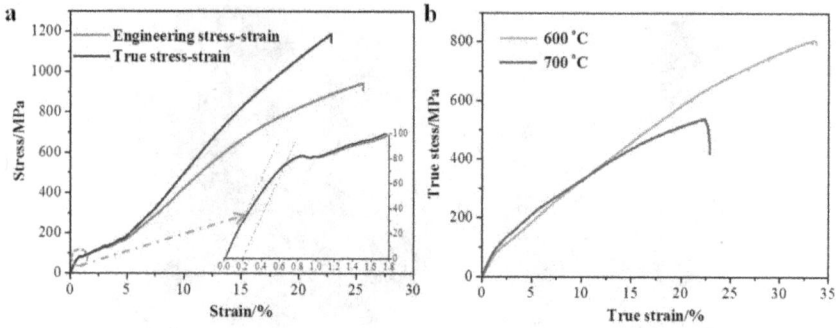

FIGURE 4.14 Tensile stress-strain curves of the AlCoCrFeNi$_{2.1}$ EHEAs: (a) room-temperature curves and (b) true tensile stress-strain curves tested at 600°C and 700°C, separately.

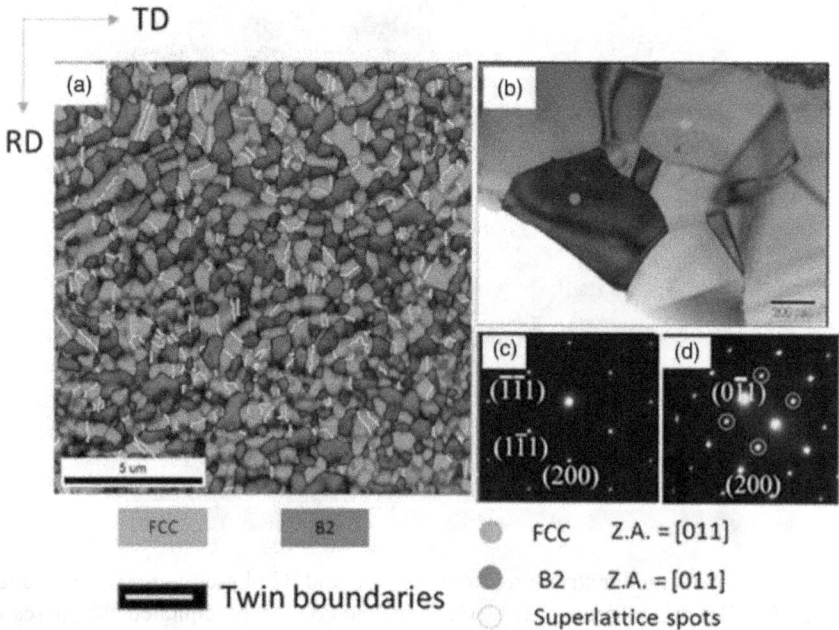

FIGURE 4.15 Microstructures of AlCoCrFeNi$_{2.1}$ EHEA after 90% cold-rolled and subsequently annealed at 800°C for 1 hour: (a) EBSD phase map, and (b) and (c) TEM bright field image and corresponding SADPs.

Bauschinger effect. Furthermore, the back-stress (BS) and effect-stress (ES) were effectively detected and displayed in the inset of Figure 4.18(b). The apparent high BS mainly contributed to the flow stress. Thus, the appealing mechanical performance ascribed to such back-stress induced effect.

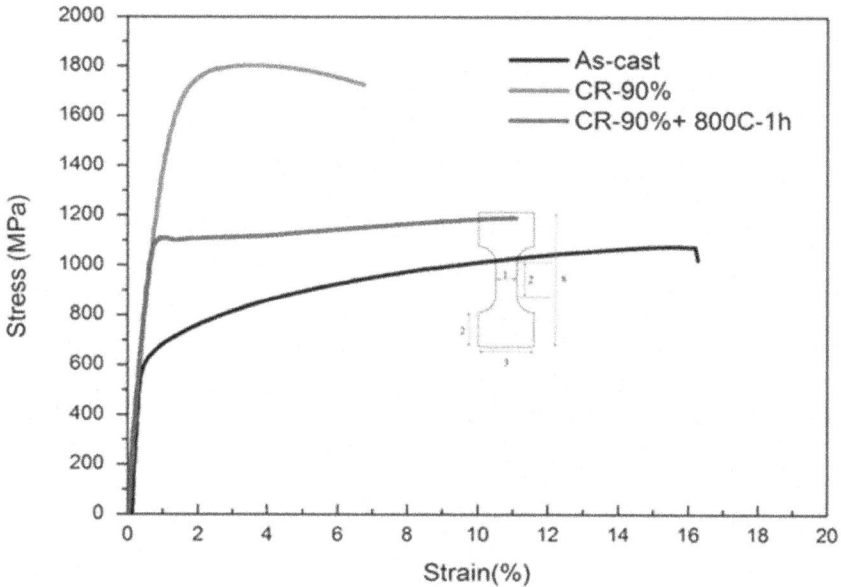

FIGURE 4.16 Engineering stress-strain curves of the AlCoCrFeNi$_{2.1}$ EHEA at the as-cast, cold-rolled, and annealed conditions.

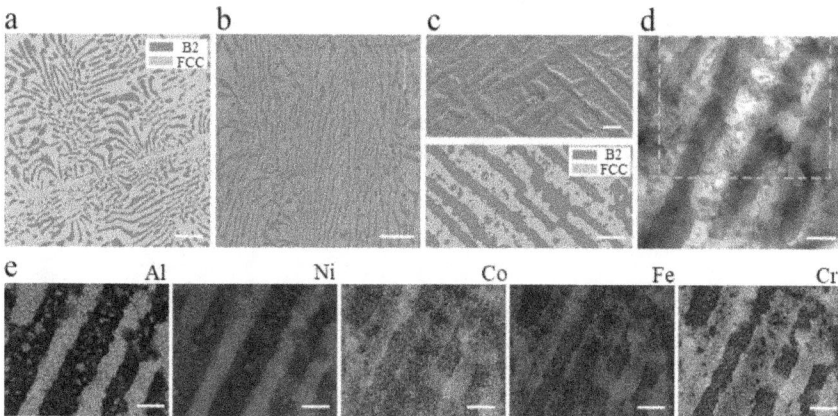

FIGURE 4.17 Microstructures of the as-cast and customized EHEA: (a) EBSD phase map of the as-cast EHEA; (b, c) SEM image and EBSD-phase image of the customized EHEA; (d) scanning TEM image exhibiting equiaxed grains in this structure; and (e) EDS maps showing the distribution of Al, Ni, Co, Fe, and Cr elements.

The mechanical behaviors under extreme conditions, such as cryogenic temperatures, are quite important to better understand the corresponding mechanical response and apply at such ultralow temperatures. Li et al. [16] investigated the mechanical performances at liquid nitrogen temperature (77 K) and unveiled the appealing

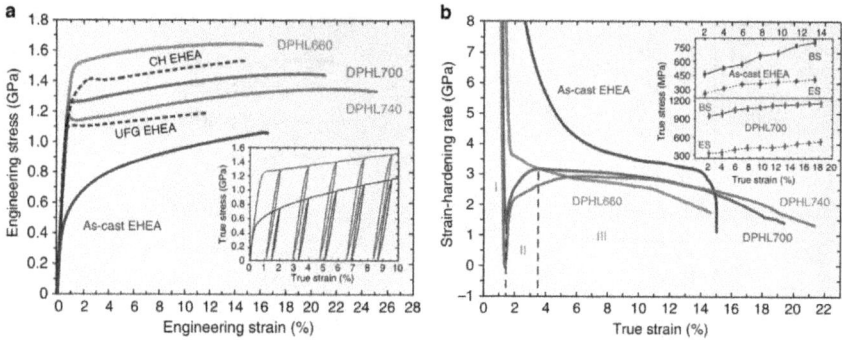

FIGURE 4.18 Mechanical responses of the three EHEAs at room temperature. (a) Tensile properties. The inset is the loading-unloading-reloading behavior of the EHEAs. (b) Strain-hardening response.

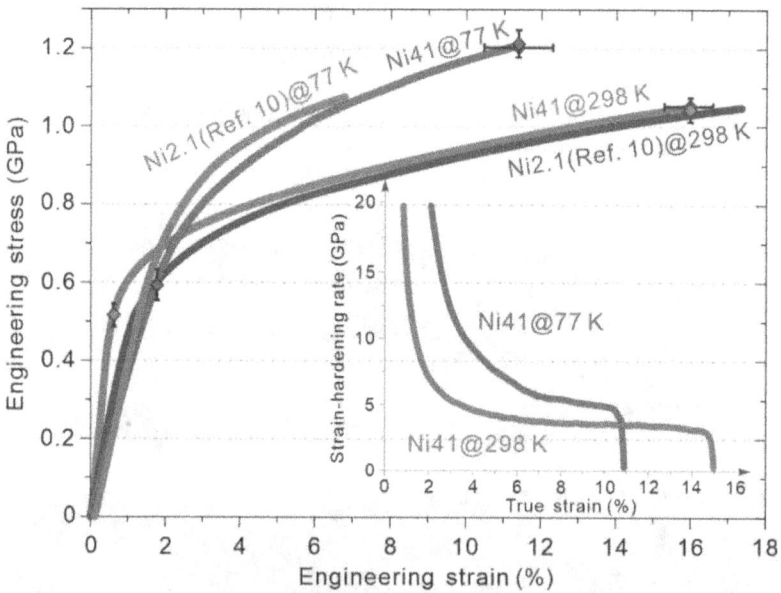

FIGURE 4.19 Tensile response curves of the as-cast $Al_{19}Co_{20}Fe_{20}Ni_{41}$ EHEA (named as Ni_{41}) and $AlCoCrFeNi_{2.1}$ EHEA ($Ni_{2.1}$) at 77 and 298 K. Inset showing the strain-hardening curves of the $Al_{19}Co_{20}Fe_{20}Ni_{41}$ EHEA at 77 and 298 K, separately.

combination of strength and ductility as well as their microstructure origins. As presented in Figure 4.19, the strengths of both $AlCoCrFeNi_{2.1}$ and $Al_{19}Co_{20}Fe_{20}Ni_{41}$ EHEAs improve; in the meantime, the ductility inevitably decreases. This case was also detected by Lu et al. [17], from which such a variation trend of properties is

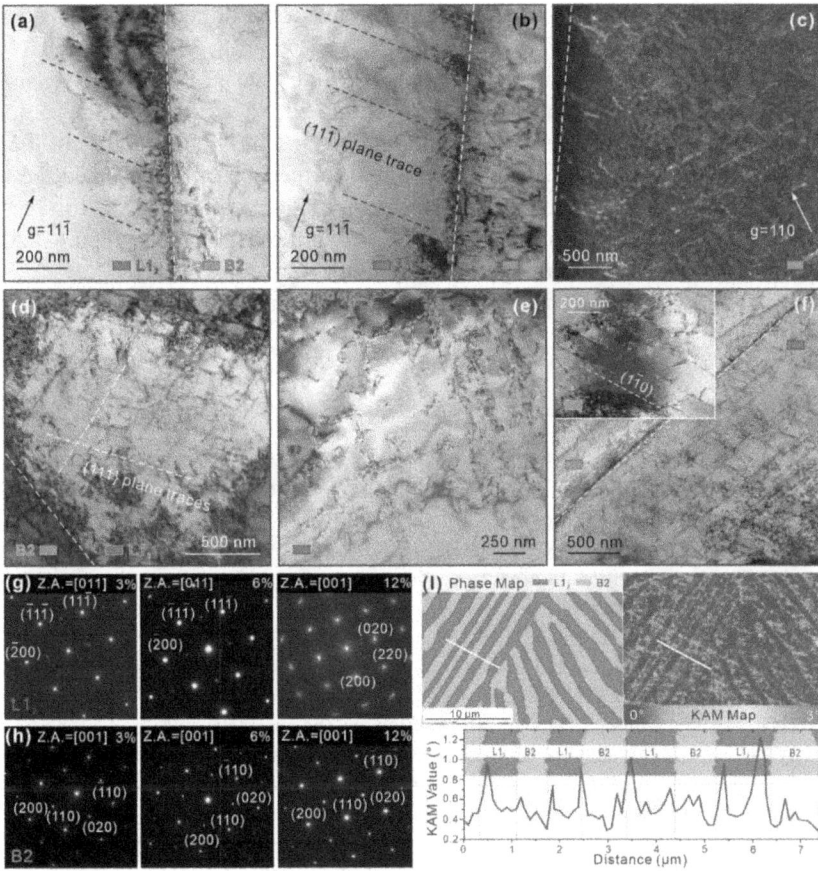

FIGURE 4.20 Deformation microstructures of the $Al_{19}Co_{20}Fe_{20}Ni_{41}$ EHEA under different strain conditions: (a) ~3%, (b, c) ~6%, and (d, e) ~12% at 77 K.

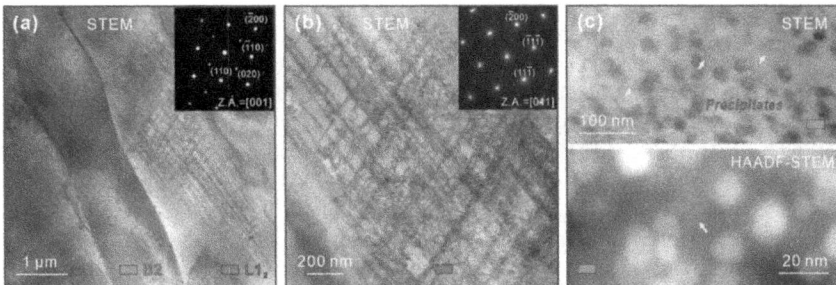

FIGURE 4.21 Deformation microstructures of the $AlCoCrFeNi_{2.1}$ EHEA at 77 K. (a, b) STEM images showing a multi-slip dislocation in the L12 phase. The insets are the SAED patterns of the B2 and L12 phases, respectively. (c) Upper and under images showing STEM and HAADF-STEM results of dense Cr-rich precipitates.

derived wherein the motion of dislocations became more difficult with temperature decrease.

With the increase in tensile deformation (~3%–~6%) at 77 K, dislocations in the L12 phase became density and intensive, featured with a series of parallel dislocations (geometrically necessary dislocations, GNDs) along the primary slip bands (Figure 4.20(a) and Figure 4.20(b)). Such GNDs formed a back-stress-induced effect and reinforced the L12 lamella. Correspondingly, forward stress formed in the B2 lamella (Figure 4.20(c)). Multiple slips activated in the L12 lamella (Figure 4.20(d) and Figure 4.20(e)) at further increase of strain to ~12%.

Similar multiple slips formed in the L12 lamella (Figure 4.21(a) and Figure 4.21(b)) in the failure specimens after tension.

REFERENCES

[1]   D. Li, C. Li, T. Feng, Y. Zhang, G. Sha, J.J. Lewandowski, P.K. Liaw, Y. Zhang, High-entropy $Al_{0.3}$CoCrFeNi alloy fibers with high tensile strength and ductility at ambient and cryogenic temperatures. *Acta Mater.* 123 (2017) 285–294.

[2]   H. Huang, Y. Wu, J. He, H. Wang, X. Liu, K. An, W. Wu, Z. Lu, Phase-transformation ductilization of brittle high-entropy alloys via metastability engineering. *Adv Mater.* 29(30) (2017) 1701678.

[3]   J.Y. He, W.H. Liu, H. Wang, Y. Wu, X.J. Liu, T.G. Nieh, Z.P. Lu, Effects of Al addition on structural evolution and tensile properties of the FeCoNiCrMn high-entropy alloy system. *Acta Mater.* 62 (2014) 105–113.

[4]   S.G. Ma, Y. Zhang, Effect of Nb addition on the microstructure and properties of AlCoCrFeNi high-entropy alloy. *Mater Sci Eng A.* 532 (2012) 480–486.

[5]   X. Yang, Y. Zhang, Prediction of high-entropy stabilized solid-solution in multi-component alloys. *Mater Chem Phys.* 132(2–3) (2012) 233–238.

[6]   Y. Zhang, T. Zuo, Y. Cheng, P.K. Liaw, High-entropy alloys with high saturation magnetization, electrical resistivity, and malleability. *Sci Rep-UK.* 3(1) (2013).

[7]   Z. Li, K.G. Pradeep, Y. Deng, D. Raabe, C.C. Tasan, Metastable high-entropy dual-phase alloys overcome the strength–ductility trade-off. *Nature.* 534(7606) (2016) 227–230.

[8]   S.S. Nene, M. Frank, K. Liu, R.S. Mishra, B.A. McWilliams, K.C. Cho, Extremely high strength and work hardening ability in a metastable high entropy alloy. *Sci Rep-UK.* 8(1) (2018).

[9]   R.X. Li, P.K. Liaw, Y. Zhang, Synthesis of Al_xCoCrFeNi high-entropy alloys by high-gravity combustion from oxides. *Mater Sci Eng A.* 707 (2017) 668–673.

[10]   S.Q. Xia, M.C. Gao, Y. Zhang, Abnormal temperature dependence of impact toughness in Al_xCoCrFeNi system high entropy alloys. *Mater Chem Phys.* 210 (2018) 213–221.

[11]   S.Q. Xia, X. Yang, T.F. Yang, S. Liu, Y. Zhang, Irradiation resistance in Al_xCoCrFeNi high entropy alloys. *JOM-US.* 67(10) (2015) 2340–2344.

[12]   S. Xia, M.C. Gao, T. Yang, P.K. Liaw, Y. Zhang, Phase stability and microstructures of high entropy alloys ion irradiated to high doses. *J. Nucl, Mater.* 480 (2016) 100–108.

[13]   Y. Lu, Y. Dong, S. Guo, L. Jiang, H. Kang, T. Wang, B. Wen, Z. Wang, J. Jie, Z. Cao, H. Ruan, T. Li, A promising new class of high-temperature alloys: eutectic high-entropy alloys. *Sci Rep-UK.* 4(1) (2015).

[14]   T.B.S.S. I. S. Wani, Ultrafine-grained $AlCoCrFeNi_{2.1}$ eutectic high entropy alloy. *Mater Res Lett.* (1160451) (2016) 174–179.

[15] P. Shi, W. Ren, T. Zheng, Z. Ren, X. Hou, J. Peng, P. Hu, Y. Gao, Y. Zhong, P.K. Liaw, Enhanced strength–ductility synergy in ultrafine grained eutectic high-entropy alloys by inheriting microstructural lamellae. *Nat Commun.* 10(1) (2019).

[16] Y. Li, P. Shi, M. Wang, Y. Yang, Y. Wang, Y. Li, Y. Wen, W. Ren, N. Min, Y. Chen, Y. Guo, Z. Shen, T. Zheng, N. Liang, W. Lu, P.K. Liaw, Y. Zhong, Y. Zhu, Unveiling microstructural origins of the balanced strength–ductility combination in eutectic high entropy alloys at cryogenic temperatures. *Mater Res Lett.* 10(9) (2022) 602–610.

[17] Y. Lu, X. Gao, L. Jiang, Z. Chen, T. Wang, J. Jie, H. Kang, Y. Zhang, S. Guo, H. Ruan, Y. Zhao, Z. Cao, T. Li, Directly cast bulk eutectic and near-eutectic high entropy alloys with balanced strength and ductility in a wide temperature range. *Acta Mater.* 124 (2017) 143–150.

5 High-Entropy Films and Coating

Yong Zhang and Xuehui Yan

5.1 INTRODUCTION

This chapter will introduce the processing of high-entropy films and coating (e.g., sputtering, laser cladding); the properties of high-entropy films; and the compositional gradient films. Specifically, we list several representative alloy thin film systems, such as AlCrFeNiTiN$_x$ films, NbMoTaWN$_x$ films, and WTaFeCrV films. In addition to introducing the traditional preparation process, we also focus on the expanded application of film preparation technology. For example, (i) utilizing multi-target co-sputtering to achieve the parallel preparation of high-entropy materials. The composition of high-entropy material was complex and diversified. Parallel preparation can provide a platform for high-throughput screening. (ii) Based on traditional sputtering technology, high-entropy flexible materials with wrinkled structures were successfully designed.

5.2 THE DEFINITION OF HIGH-ENTROPY FILMS

High-entropy films (HEFs) show the same scientific concept as high-entropy alloys, which contain multiple components and possess a concentrated solid-solution structure [1–5]. At present, alloy thin films are widely used in industrial fields, such as protective coatings, wear-resistant coatings, and thermal barrier coatings [6–10]. In addition to alloy films, traditional industrial films are also classified into nitride films [11,12], carbide films, and oxide films [10]. In contrast with traditional films, HEFs exhibit excellent properties similar to those of high-entropy alloys and even outperform the bulk alloys in some properties [13].

For composition design, there are about 25 kinds of elements that can be used for the design of HEFs, which are mainly concentrated in the transition metal region. In general, the elements used for the composition design of HEFs can be divided into three categories—base elements, functional elements, and non-metal elements: (i) for the base elements, it generally refers to the Cr, Fe, Co, Ni, Cu elements, which have similar atomic sizes and tend to form a simple face-centered cubic (FCC) or body-centered cubic (BCC) solid-solution structure; (ii) for functional elements, it generally refers to Ti, Mn, V, W, and other elements, which possess excellent heat resistance, corrosion resistance, and other special properties; (iii) for the non-metal elements, it generally refers to some small-sized non-metallic elements, such as B, C, N, and O, which can fill the interstitial positions of the solid-solution structure and further improve hardness films.

DOI: 10.1201/9781003319986-5

5.3   THE PREPARATION METHODS OF HIGH-ENTROPY FILMS

The preparation methods of HEFs are mainly divided into two categories, including physical vapor deposition (PVD) and chemical vapor deposition (CVD). Among them, the PVD methods are the most common method to prepare HEFs, such as vacuum sputtering, vacuum evaporation, and ion plating. More precisely, magnetron sputtering and laser cladding techniques are the most commonly used preparation methods for industrial HEFs [14–18]. Magnetron sputtering technology uses the sputtering effect. High-energy particles bombard the surface of the target, and the target atoms escape and move in a specific direction and, finally, deposit on the substrate to form a thin film. The dual function of magnetic and electric fields increases the collision probabilities of the electron, charged particles, and gas molecules. For the reactive sputtering, the deposition atmosphere is a mixed state, such as argon and nitrogen, and argon and oxygen. In this case, the metal atoms reacted with N/O/C atoms and were deposited on the substrates. The schematic diagram of reactive sputtering is shown in Figure 5.1 [19].

For laser cladding techniques, it is pointed out that melting metal powder features certain physical, chemical, or mechanical properties by high-power and high-speed laser. A layer combined with the matrix in the way of metallurgy bonding can improve the mechanical properties between the layer and matrix. Laser cladding technology is divided into two types of methods, referred to as pre-powder and synchronous feeding, as shown in the schematic diagram of Figure 5.2 [1]. In contrast with magnetron sputtering, the thickness of the coating prepared by laser cladding is

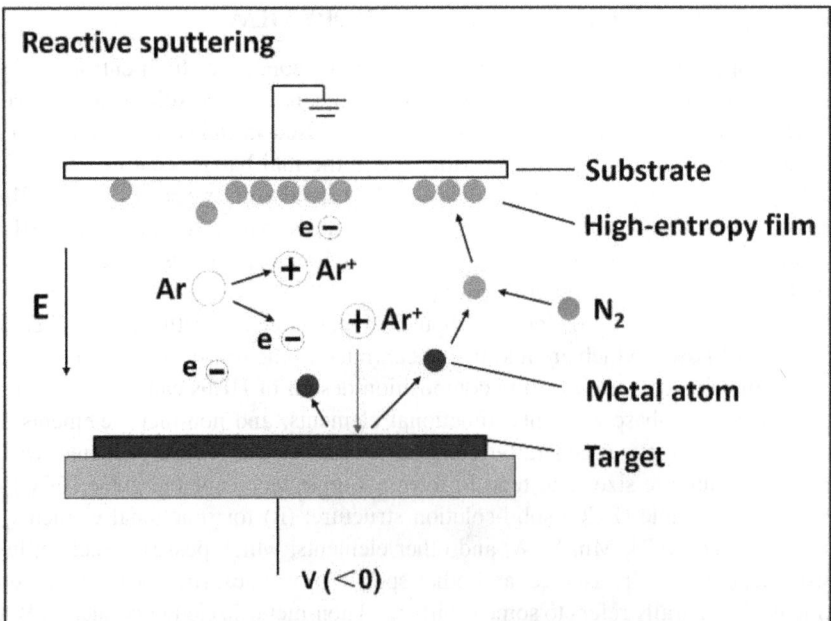

FIGURE 5.1   The schematic diagram of reactive sputtering [19].

a

Reflector

Laser

Powder nozzle

Coating

Substrate

Work direction

b

Reflector

Laser

Preset powder layer

Coating

Substrate

Work direction

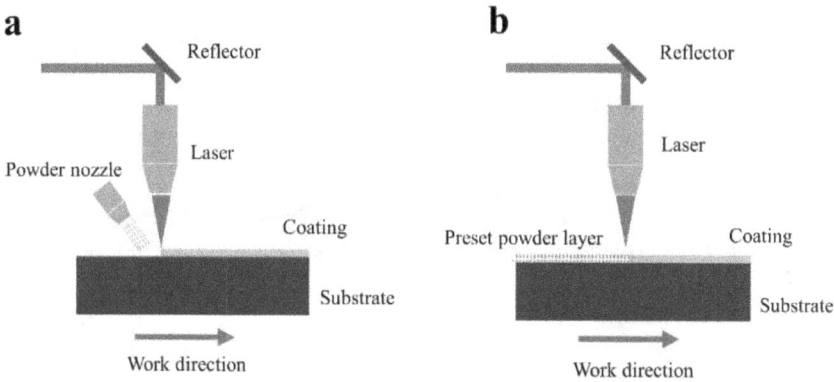

FIGURE 5.2 Schematic of laser cladding: (a) synchronous powder and (b) pre-powder layer [1].

larger than films prepared by sputtering. Moreover, the bonding strength is stronger than films prepared by magnetron sputtering, which is attributed to the coating and substrate being metallurgically bonded. There are also several unique characteristics of magnetron techniques, such as (i) no special requirement for the conductivity of the target, (ii) good compositional consistency and structural stability, and (iii) flexible control of film properties and thickness.

5.4 SEVERAL TYPICAL HIGH-ENTROPY FILMS

5.4.1 HIGH-HARDNESS FILMS: MICROSTRUCTURE AND PROPERTIES OF ALCRFENITIN$_x$ FILMS

Zhang and co-workers have reported on $(Al_{0.5}CrFeNiTi_{0.25})N_x$ films prepared by reactive sputtering [19]. In this work, the ratio of $N_2/(Ar + N_2)$ is set as a variable. With the increase of the N content, the phase structures of $(Al_{0.5}CrFeNiTi_{0.25})N_x$ films transform from amorphous to FCC phase structure, as shown in Figure 5.3. Moreover, it is noteworthy that the bulk $Al_{0.5}CrFeNiTi_{0.25}$ alloys featured a BCC structure, which is totally different from nitride alloy films. Due to the high cooling rate, films are far from the equilibrium that could be achieved by bulk alloys, which cause an amorphous structure. As N content increases, a concentrated solid-solution structure is formed rather than complex multi-nitride phases. It should be noted that nitrides like TiN, CrN, AlN, and FeN also possess a simple FCC structure. It can be inferred that the solid solution occurred between nitrides in HEFs, as shown in Figure 5.4.

For the mechanical properties, the HEFs showed the highest hardness and Young's modulus of 21.78 GPa and 253.8 GPa. Compared with the bulk $Al_{0.5}CrFeNiTi_{0.25}$ alloys (the value of hardness was about 6 GPa), the hardness of the HEFs was improved significantly. More significantly, the hardness of HEFs is also enhanced with the increase of the N content. It proved that the interstitial atoms could significantly increase the hardness of HEFs. By adding N and other small-size elements, the hardness of the HEFs can be further enhanced and show a rising trend than

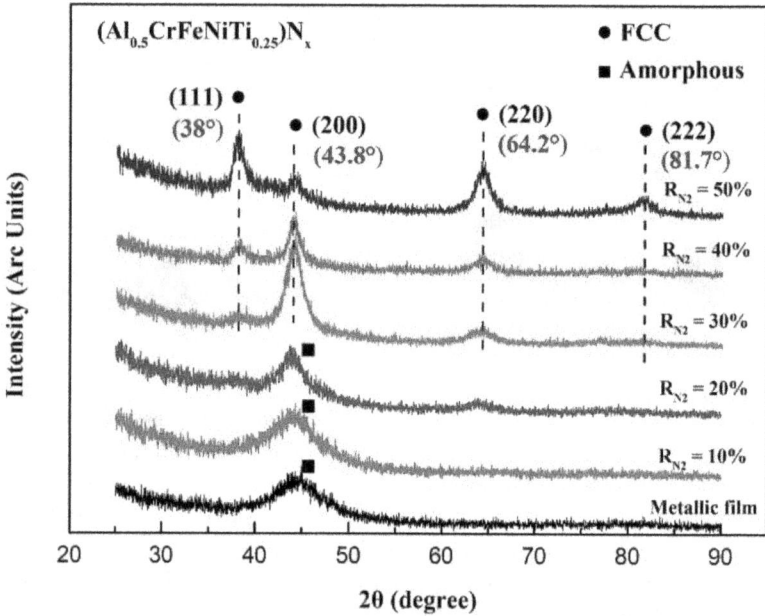

FIGURE 5.3 X-ray patterns of the $(Al_{0.5}CrFeNiTi_{0.25})N_x$ films deposited at different R_{N2} [19].

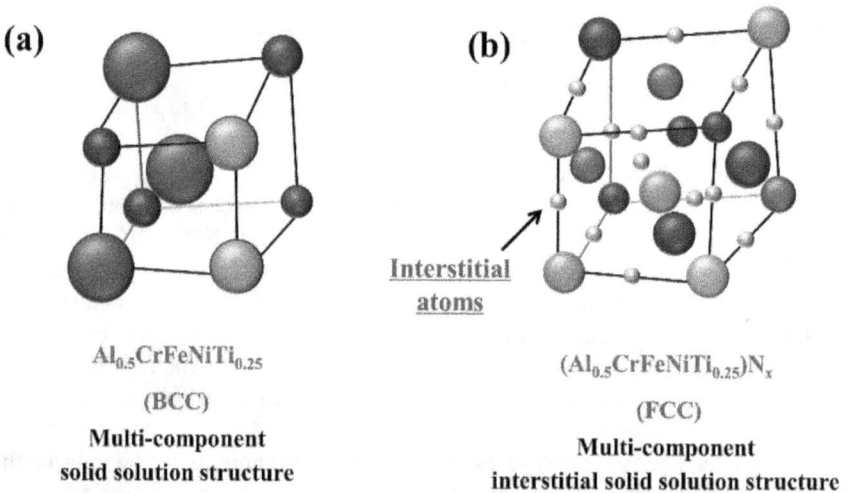

FIGURE 5.4 Schematic of the lattice structure: (a) $Al_{0.5}CrFeNiTi_{0.25}$ bulk and (b) $(Al_{0.5}CrFeNiTi_{0.25})N_x$ high-entropy films [19].

alloy films. This was attributed to the fact that solid solution significantly increased the hardness. In this case, industrial hard coatings and wear-resistant coatings are generally used in ceramic films. HEFs can be developed as a potential wear-resistant coating due to its concentrated solid-solution structure and severe lattice distortion.

5.4.2 Thermal Stability Films: Microstructure and Properties of NbTiAlSiW$_x$N$_y$ Films

Zhang and co-worker have also designed NbTiAlSiWN$_x$ films by reactive sputtering in the atmosphere of a mixture of N$_2$ + Ar, which shows excellent thermal stability [20]. The NbTiAlSiW$_x$N$_y$ films ($x = 0$ or 1) possess an amorphous structure. More excitingly, the films show a beautiful color, and the color is changed with different thicknesses, as shown in Figure 5.5. The color is sensitive to this property; commonly, the thicker films displayed darker colorations, while the addition of nitrogen produced lighter-colored films. Due to the unique features, the NbTiAlSiW$_x$N$_y$ films are expected to be developed as a decorative coating.

Moreover, the NbTiAlSiW$_{x-}$N$_y$ films show excellent thermal stability. For all of the as-deposited NbTiAlSiWN$_y$ films produced using various N$_2$ flow rates, it is clear that only one broad peak exists. This indicates the as-deposited NbTiAlSiWN$_y$ films have an amorphous structure. Sluggish diffusion effects, high mixing entropy, and severe lattice distortions all contribute to the formation of an amorphous structure. All the as-deposited films present amorphous structures, which remain stable at 700°C for over 24 h, as shown in Figure 5.6.

This indicates that the NbTiAlSiW$_x$N$_y$ films could be good candidates as a protective refined coating. Moreover, HEFs with good high-temperature stability have also shown great development potential in photothermal conversion films [21]. Through the test of NbTiAlSiW$_x$N$_y$ thin films, it is found that the light absorption rate of alloy films can reach to about 80%. The optimization of multilayer films in the field of photothermal conversion coatings still requires extensive experiments.

FIGURE 5.5 The macrophotograph of the deposited thin films of different colors [20].

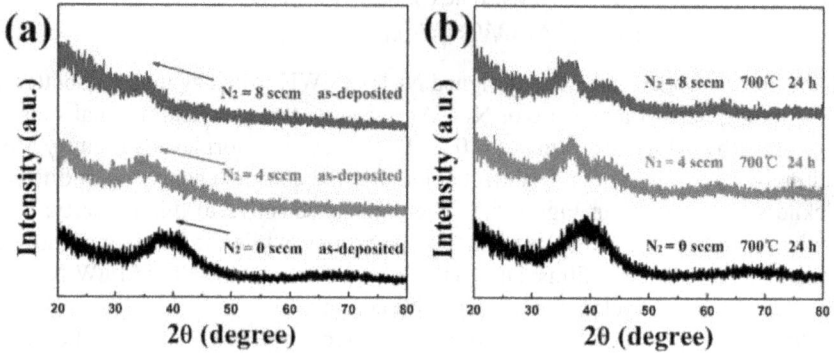

FIGURE 5.6 XRD patterns of NbTiAlSiWN$_x$ film before and after heat treatment: (a) as-deposited and (b) 700°C for 24 h [20].

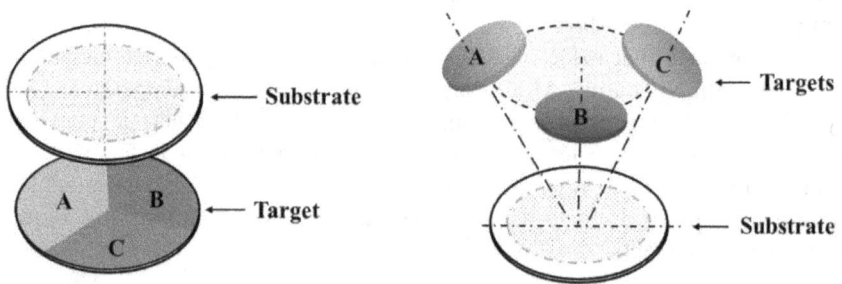

FIGURE 5.7 Schematic of multi-target co-deposition [1].

5.4.3 COMPOSITIONAL GRADIENT FILMS

Multi-target co-sputtering technology is an effective way to realize the preparation of gradient thin films [22–24]. Based on the composition gradient film, the composition, structure, and properties of the alloy can be screened rapidly. As shown in Figure 5.7, during sputtering, the target atoms leave the target surface in a simple cosine distribution, and the target atoms are deposited toward the substrate at a specific angle and direction. Using the relative spatial distance difference between the substrate and each target, the deposition density difference in the horizontal direction is obtained, and finally, each component is deposited on the substrate in a continuous gradient.

Zhang et al. have designed a pseudo-ternary alloy system of TiAlCrFeNi. Utilizing the three targets for co-sputtering, the parallel preparation is achieved, and 144 alloy units are obtained at one time (Figure 5.8 and Figure 5.9) [25]. Three targets were 120° apart and focused on a stable silicon wafer. The material library shows large composition coverage. The atomic fraction of Al varies from 9.06% to 89.25%, Ti is 3.34% to 84.39%, and FeCrNi is 4.36% to 79.61%. Using the multi-target co-sputtering method to prepare the composition gradient material, the continuous composition gradient

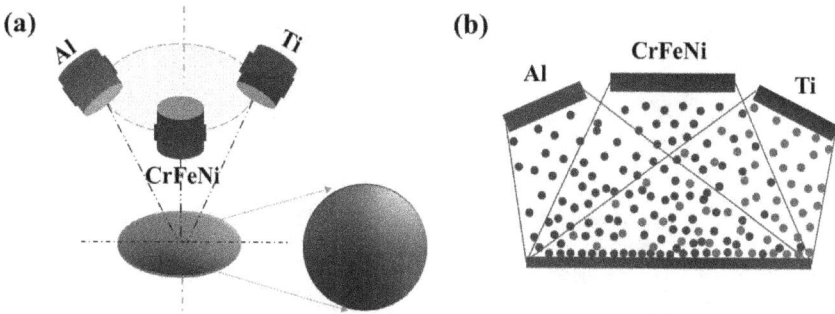

FIGURE 5.8 Schematic diagram of multitarget co-sputtering: (a) three-dimensional view co-sputtering and (b) front view and the process of deposition [25].

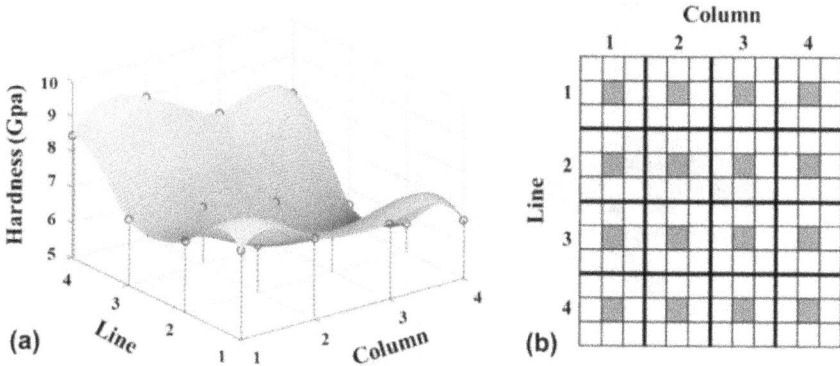

FIGURE 5.9 (a) Nanoindetation hardness of TiAlCrFeNi alloy system; (b) location schematic diagram of selected sample units [25].

change can be obtained without the assistance of any template, the obtained film has a uniform composition in the thickness direction, and the deposited film does not require heat treatment to diffuse. Using multi-target co-sputtering deposition, the change law of the composition gradient is nonlinear; that is, the composition gradient per unit distance is different at different positions of the substrate. Therefore, the composition gradient cannot be precisely controlled. A preliminary screening of the phase structure and properties were performed. The phases are mainly composed of an amorphous phase and a body-centered cubic phase. Hardness changes nonlinearly with compositions. The material library synthesized in this study is expected to provide an effective platform for the high-throughput screening of multi-component materials.

The mechanical properties of this material library are also tested by nanoindentation. To ensure representativeness, the material library was divided into nine parts, and the central sample unit of each part was taken for the characterization of mechanical properties. The hardness of the sample units and location are shown in Figure 5.10.

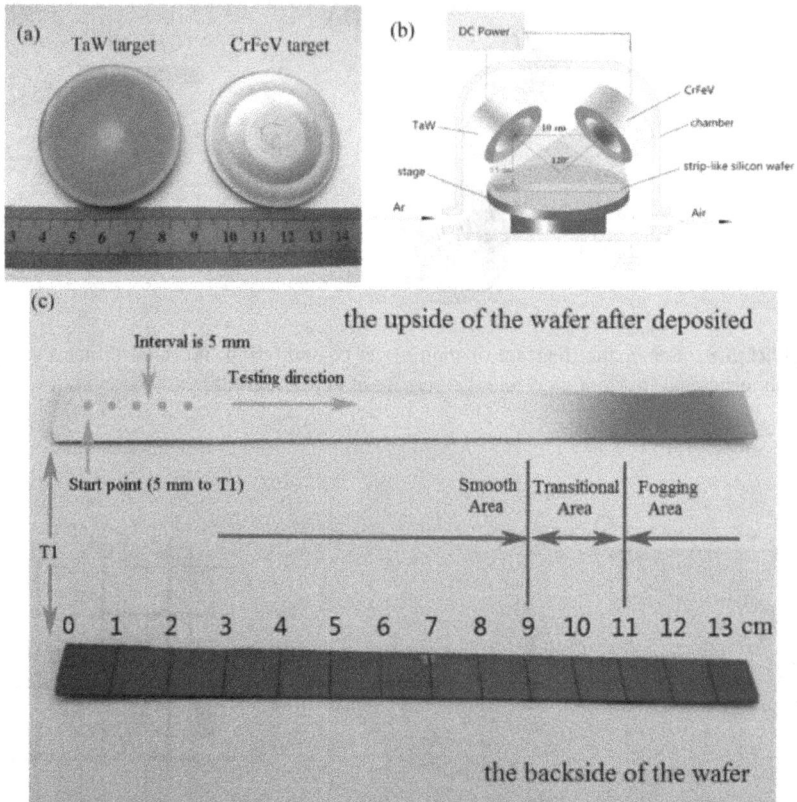

FIGURE 5.10 (a) TaW and CrFeV targets used in the present work. (b) Magnetron sputtering process to deposit (Cr, Fe, V)–(Ta, W) high-entropy film. (c) The upside of the wafer after deposition (top) and the backside of the wafer was marked before deposition (bottom). Orange points and arrows indicate test points and selected directions on the wafer [26].

The trend of hardness can give a direct guide to screening the alloys with excellent mechanical properties. The material library synthesized in this method is stable and homogenous and exhibits large coverage in various compositions. High-throughput technology is urgently needed for multiple component materials, and the synthesis of materials with composition gradients is the key step for high-throughput screening.

Similarly, the five-element high-entropy alloy system was designed as a pseudo-binary, namely (Ta, W)–(Cr, Fe, V), and alloy targets were prepared, respectively, and double-target co-sputtering was designed [26]. By utilizing the spatial gradient during deposition and sputtering, a non-uniform deposition density is obtained, and finally, a continuously changing composition gradient is obtained. A total of 13 sample units were isolated in a single preparation, and their physical map is shown in Figure 5.11. By verifying points in the compositional library, the structure and property variation according to the atomic content of elements are carefully studied. Results indicate that the films exhibit an amorphous structure when x ranges from

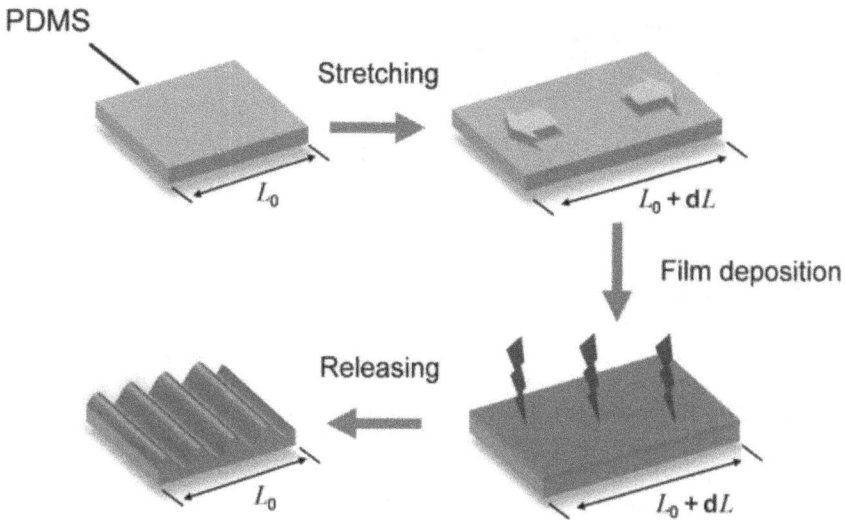

FIGURE 5.11 Schematic of the wrinkled-structure fabrication process [27].

86.9 to 32.5, and high concentrations of Ta and W lead to the formation of a BCC structure in the films.

5.4.4 FLEXIBLE WRINKLE-STRUCTURED FILMS

A controlled wrinkled structure is a simple and effective approach to achieving unique properties and has been widely used in flexible materials. Huang et al. reported a substrate prestrain method for fabricating wrinkle-structured $Zr_{52}Ti_{34}Nb_{14}$ multiple-basis-element (MBE) alloy films as biocompatible materials [27]. The preparation of high-entropy flexible thin films is realized based on magnetron sputtering technology, and $Zr_{52}Ti_{34}Nb_{14}$ thin films are prepared with surface-wrinkled microstructures on PDMS substrates by using a stretching device and substrate prestraining method, such as shown in Figure 5.11.

By controlling the film thickness and the size of the prestrain, the scale of the wrinkled structure was adjusted from micron to nanometer, as shown in Figure 5.12. At certain prestrains, it is clearly observed that the morphology of the wrinkled structure and the size of the single wrinkle increased with the change in the film thickness, which proves that the structure can be effectively tuned by the thickness changes. Moreover, they have investigated the different prestrains on the wrinkled structure, as shown in Figure 5.13. Results showed that there is an obvious difference between samples with different prestrains.

5.5 OUTLOOK

Briefly, HEFs have shown many excellent properties, such as good thermal stability, excellent corrosion resistance, ultrahardness, and outstanding wear resistance. HEFs

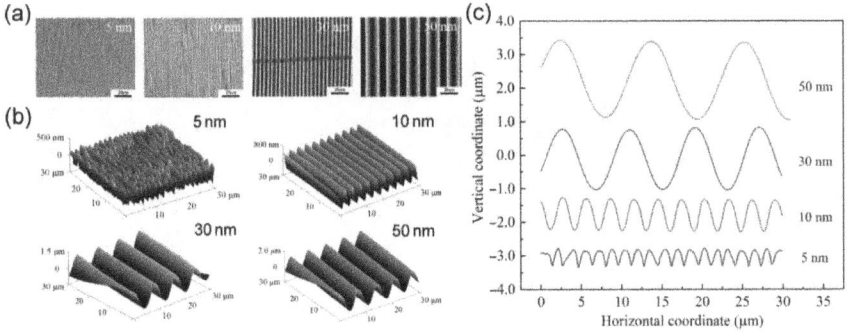

FIGURE 5.12 Wrinkled structure of $Zr_{52}Ti_{34}Nb_{14}$ films with different thicknesses: (a) optical morphology, (b) 3D morphology, and (c) cross-section [27].

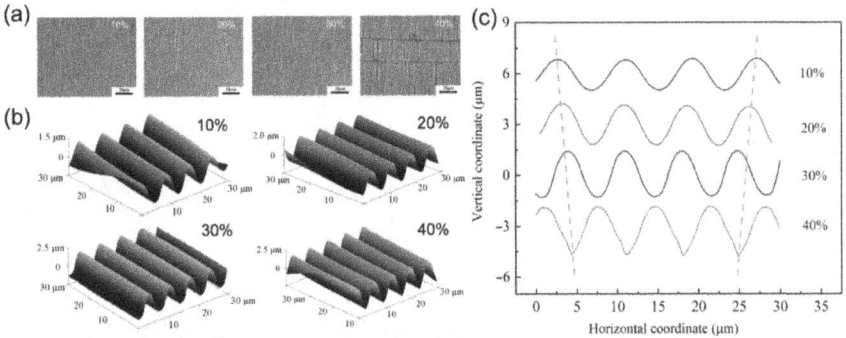

FIGURE 5.13 Wrinkled structure of $Zr_{52}Ti_{34}Nb_{14}$ films with different prestrains: (a) optical morphology, (b) 3D morphology, and (c) cross-section [27].

have become potential films and coatings in wide industrial applications. However, the composition design is more complex than that of traditional coatings. In this case, the mechanism research on HEFs is more difficult due to the complex structures. It is undeniable that there is still a long way to further in achieving its industrial application.

Moreover, many efforts conducted on HEFs are mainly focused on mechanical properties and physical properties, such as hardness, corrosion resistance, and oxidation resistance. Research is relatively weak on the phase formation of HEFs. Due to the high cooling rate of HEAs, HEFs eventually fail to reach the equilibrium state that is similar to the bulk state. Hence, the phase formation rules for bulk alloys will no longer apply to HEFs. It is necessary to establish the phase formation law of the HEFs, which will help to achieve direct guidance on the structural design of HEA thin films. Furthermore, based on the common preparation method of HEFs, compositional gradient films have also been developed. More excitingly, the compositional gradient films can be used as a material library to achieve the screen of high-entropy materials. Due to the characterization of multi-principal components in high-entropy materials, the cost of high-entropy alloys is inevitably higher than traditional alloys. For typical industrial applications, this is a shortcoming. In this case, developing high-entropy films and coatings as a special protective coating is significant to realize their industrial applications.

REFERENCES

[1] Yan X H, Li J S, Zhang W R, et al. A brief review of high-entropy films[J]. *Materials Chemistry and Physics*, 2018, 210: 12–19.

[2] Zhang Y, Zuo T T, Tang Z, et al. Microstructures and properties of high-entropy alloys[J]. *Progress in Materials Science*, 2014, 61(Apr.): 1–93.

[3] Yan X, Zhang Y. Functional properties and promising applications of high entropy alloys[J]. *Scripta Materialia*, 2020, 187: 188–193.

[4] Gao M C, Miracle D B, Maurice D, et al. High-entropy functional materials[J]. *Journal of Materials Research*, 2018, 33(19): 3138–3155.

[5] Yeh J W, Chen S K, Lin S J, et al. Nanostructured high-entropy alloys with multiple principal elements: novel alloy design concepts and outcomes[J]. *Advanced Engineering Materials*, 2004, 6(5): 299–303.

[6] Chen T-K, Wong M-S, Shun T-T, et al. Nanostructured nitride films of multi-element high-entropy alloys by reactive DC sputtering[J]. *Surface and Coatings Technology*, 2005, 200(5): 1361–1365.

[7] Huang Y-S, Chen L, Lui H-W, et al. Microstructure, hardness, resistivity and thermal stability of sputtered oxide films of AlCoCrCu0.5NiFe high-entropy alloy[J]. *Materials Science and Engineering: A*, 2007, 457(1–2): 77–83.

[8] Tsai M-H, Yeh J-W, Gan J-Y. Diffusion barrier properties of AlMoNbSiTaTiVZr high-entropy alloy layer between copper and silicon[J]. *Thin Solid Films*, 2008, 516(16): 5527–5530.

[9] Lin D-Y, Zhang N-N, He B, et al. Tribological properties of FeCoCrNiAlBx high-entropy alloys coating prepared by laser cladding[J]. *Journal of Iron and Steel Research International*, 2017, 24(2): 184–189.

[10] Ren B, Liu Z, Li D, et al. Corrosion behavior of CuCrFeNiMn high entropy alloy system in 1 M sulfuric acid solution[J]. *Materials and Corrosion*, 2012, 63(9): 828–834.

[11] Shen W-J, Tsai M-H, Chang Y-S, et al. Effects of substrate bias on the structure and mechanical properties of (Al1.5CrNb0.5Si0.5Ti) Nx coatings[J]. *Thin Solid Films*, 2012, 520(19): 6183–6188.

[12] Pogrebnjak A D. Structure and properties of nanostructured (Ti-Hf-Zr-V-Nb) N coatings[J]. *Journal of Nanomaterials*, 2013, 2013.

[13] Chang Z-C, Liang S-C, Han S, et al. Characteristics of TiVCrAlZr multi-element nitride films prepared by reactive sputtering[J]. *Nuclear Instruments and Methods in Physics Research Section B: Beam Interactions with Materials and Atoms*, 2010, 268(16): 2504–2509.

[14] Zhang H, He Y Z, Pan Y, et al. Synthesis and characterization of NiCoFeCrAl3 high entropy alloy coating by laser cladding[C]. *Advanced Materials Research*, 2010: 1408–1411.

[15] Zhang H, Pan Y, He Y. The preparation of FeCoNiCrAl2Si high entropy alloy coating by laser cladding[J]. *Journal of Metals in Chinese*, 2011, 8: 1075–1079.

[16] Huang C, Zhang Y, Shen J, et al. Thermal stability and oxidation resistance of laser clad TiVCrAlSi high entropy alloy coatings on Ti—6Al—4V alloy[J]. *Surface and Coatings Technology*, 2011, 206(6): 1389–1395.

[17] Feng X, Tang G, Sun M, et al. Structure and properties of multi-targets magnetron sputtered ZrNbTaTiW multi-elements alloy thin films[J]. *Surface and Coatings Technology*, 2013, 228: S424–S427.

[18] Liang S-C, Tsai D-C, Chang Z-C, et al. Structural and mechanical properties of multi-element (TiVCrZrHf)N coatings by reactive magnetron sputtering[J]. *Applied Surface Science*, 2011, 258(1): 399–403.

[19] Zhang Y, Yan X-H, Liao W-B, et al. Effects of nitrogen content on the structure and mechanical properties of (Al0.5CrFeNiTi0.25)Nx high-entropy films by reactive sputtering[J]. *Entropy*, 2018, 20(9): 624.

[20] Sheng W, Yang X, Wang C, et al. Nano-crystallization of high-entropy amorphous NbTiAlSiWxNy films prepared by magnetron sputtering[J]. *Entropy*, 2016, 18(6): 226.

[21] Sheng W-J, Yang X, Zhu J, et al. Amorphous phase stability of NbTiAlSiNX high-entropy films[J]. *Rare Metals*, 2018, 37(8): 682–689.

[22] Ding S, Liu Y, Li Y, et al. Combinatorial development of bulk metallic glasses[J]. *Nature Materials*, 2014, 13(5): 494–500.

[23] Mao S S. High throughput growth and characterization of thin film materials[J]. *Journal of Crystal Growth*, 2013, 379: 123–130.

[24] Yan X-H, Ma J, Zhang Y. High-throughput screening for biomedical applications in a Ti-Zr-Nb alloy system through masking co-sputtering[J]. *Science China Physics, Mechanics & Astronomy*, 2019, 62(9): 996111.

[25] Zhang Y, Yan X, Ma J, et al. Compositional gradient films constructed by sputtering in a multicomponent Ti-Al-(Cr, Fe, Ni) system[J]. *Journal of Materials Research*, 2018, 33(19): 3330–3338.

[26] Xing Q, Ma J, Wang C, et al. High-throughput screening solar-thermal conversion films in a pseudobinary (Cr, Fe, V)-(Ta, W) system[J]. *ACS Combinatorial Science*, 2018, 20(11): 602–610.

[27] Huang H, Liaw P K, Zhang Y. Structure design and property of multiple-basis-element (MBE) alloys flexible films[J]. *Nano Research*, 2022, 15(6): 4837–4844.

6 High-Entropy Fibers

Yong Zhang and Ruixuan Li

6.1 INTRODUCTION

Materials can generally be divided into rigid and flexible materials. For alloy materials, bulk alloys are generally considered to be rigid materials, and fibers or thin ribbons with certain flexibility are flexible materials. Generally, alloy fibers have good electrical conductivity, thermal conductivity, high strength, high elasticity, flexibility, wear resistance, corrosion resistance, and high temperature resistance, as well as shielding, anti-magnetic, and anti-radiation effects, and so on. In particular, high-strength and high-toughness alloy fibers have important uses in key infrastructure, such as large cable-stayed bridges, suspension bridges, and special rescue equipment.

Preliminary studies have confirmed that, due to the novel compositional design concept and the complex regulation of a strengthening mechanism, bulk HEAs are strong candidates for overcoming the strength-ductility trade-off [1]. At the same time, HEAs also exhibit a series of excellent characteristics, such as radiation resistance, corrosion resistance, and good high-temperature stability [2–4]. In this case, HEA fibers become one of the important candidates for high-strength and high-toughness fiber materials. However, it should be noted that for fibers, the increase in the amount of deformation will lead to a significant reduction in its diameter. When the scale of material plastic deformation carriers, such as dislocation lines and twin defects, is in a similar order to its geometric size, it may cause the metal material to exhibit plastic deformation behavior different from that of the macroscale material, which is called the scale effect. Therefore, it is very important to study the preparation method and mechanical properties of HEA fibers. This chapter summarizes the research status of HEA fibers and gives an outlook on their future development.

6.2 HEA FIBERS PREPARED BY HOT-DRAWING

D. Li et al. took the lead in investigating the preparation and mechanical properties of HEA fibers [5]. They successfully prepared $Al_{0.3}CoCrFeNi$ fibers with diameters of 1–3.15 mm by rotary forging and hot-drawing. Rotary forging, also known as radial forging, is a forming process for long-shaft rolling stock. Its working principle is shown in Figure 6.1. The sample is radially symmetrically arranged with more than two hammer heads, which strike with high-frequency radial reciprocating motion. Then the sample rotates and moves axially, and it realizes radial

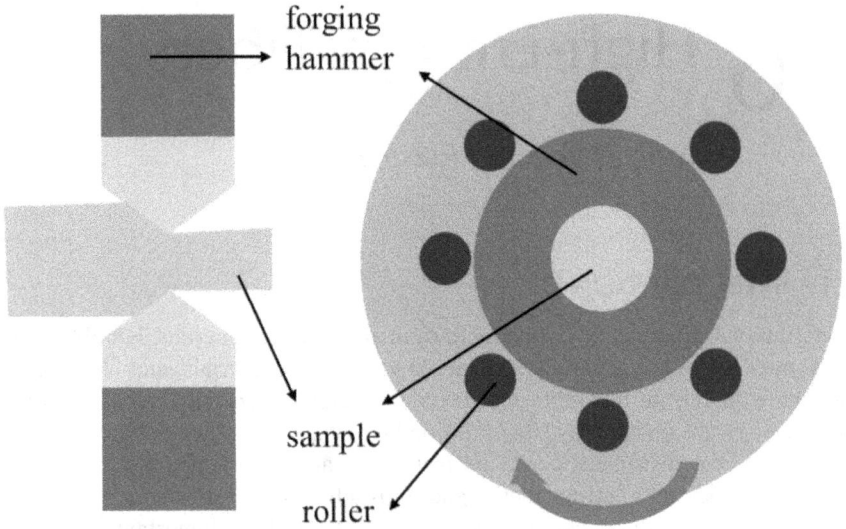

FIGURE 6.1   The working principle of rotary forging.

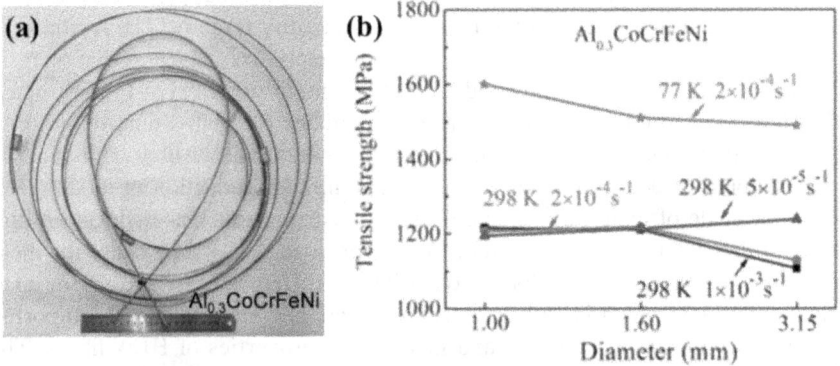

FIGURE 6.2   (a) The macroscopic morphology of the fibers and (b) tensile properties of fibers at room temperature and 77K with different diameters [5].

compression and length extension deformation. Rotary forging has a lot of blows, and the surface of each blow is very similar to rolling. As a result, it has the characteristics of pulse loading and multi-directional forging, which is beneficial to the uniform deformation of metal and the improvement of plasticity. The forgings prepared by this method are of high quality and high dimensional accuracy. The HEA fibers with good quality can be prepared by hot-drawing the forging is prepared by rotary forging.

The authors carried out detailed characterization and testing of the microstructure and mechanical properties of the HEA fibers prepared by this method. Figure 6.2

shows the macroscopic morphology of the fibers and their tensile properties at room temperature and 77 K with different diameters. They found that in the matrix, the distribution of the five constituent elements was generally uniform, but because the Al and Ni atoms had the most negative mixing enthalpy, some B2-structured Al-Ni nanoprecipitates still appeared. Precipitates can be produced in $Al_{0.3}CoCrFeNi$ HEA fibers with different diameters. The use of EBSD gave a grain size of about 1.6 μm in the fibers and the appearance of <111> and <100> textures along the axial direction. In terms of mechanical properties, the HEA fiber not only has excellent mechanical properties at room temperature but also increases its strength and plasticity at 77 K. The tensile strength is up to 1.5 GPa, and the plasticity is about 20% at 77 K, which is because of the deformation mechanism transformation from the dislocation plane-slip to the nano-twinning. This is beneficial for the low-temperature application of HEA fibers of this composition.

The authors further studied the effect of the annealing process on the microstructure and mechanical properties of $Al_{0.3}CoCrFeNi$ HEA fibers [6]. They found that in the annealed state, a large number of B2 precipitates and annealing twins were formed, and with the prolongation of annealing time, the grain size did not differ much, all less than 3 μm. Compared with the as-drawn state, the elongation after heat treatment was significantly increased, but the yield strength and breaking strength decreased.

Figure 6.3 compares ultimate tensile strength and yield strength vs. elongation at room temperature of the $Al_{0.3}CoCrFeNi$ HEA fibers and other traditional alloy fibers (including CuSn, CuAlNi, CuAl, NiMnGa, ZnLi, Cu, MgYZn, and stainless steel). It can be seen that the yield strength and tensile strength of $Al_{0.3}CoCrFeNi$ HEA fibers greatly exceed those of other traditional alloy fibers, which are mainly due to the compound reinforcement effect of HEAs, including solid-solution hardening, grain boundary hardening, precipitation hardening, and dislocation hardening.

J.P. Liu et al. used a combination of hot-forging, hot-rolling, and hot-drawing to prepare a CoCrNi alloy fiber with a diameter of 2 mm [7]. They first hot-forged the homogenized sample at 1423 K with a diameter of 15 mm and then continued to hot-roll to 8 mm at 1123 K using the caliber rolling equipment. Finally, the 8 mm rod was hot-drawn ten times at 1073 K with a drawing speed of 4 m/min.

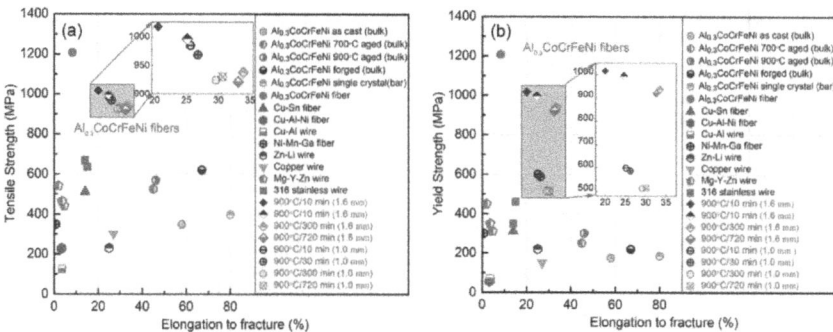

FIGURE 6.3 Comparison of mechanical properties between $Al_{0.3}CoCrFeNi$ HEA fibers and other fibers [6]. (a) Tensile strength-elongation, (b) Yield strength-elongation.

CoCrNi wire has excellent mechanical properties not only at room temperature but also at 77 K. It has a tensile strength of 1220 MPa at room temperature and an elongation at break of 24.5%. When the temperature drops to 77 K, its strength and plasticity increase at the same time, the tensile strength is 1783 MPa, and the elongation at break reaches 37.4%. This is because the CoCrNi alloy is a low stacking-fault energy alloy (the stacking-fault energy is about 18 mJ/m^{-2}), and the low stacking-fault energy can effectively promote the formation of twins. A large number of twins and dislocation walls play the role of refining the grains and hindering the movement of dislocations, improving the work-hardening ability of the fibers. At low temperatures, not only high-density nano-twins and stacking faults are maintained, but also FCC→HCP martensitic transformation occurs simultaneously, which further improves its work-hardening ability and comprehensive mechanical properties. Therefore, compared with the traditional pearlitic steel wire, this advanced CoCrNi HEA fiber is easy to prepare on the one hand and has strong engineering application potential on the other hand, especially in a low-temperature environment.

6.3 HEA FIBERS PREPARED BY COLD-DRAWING

In addition to hot-drawing to prepare alloy fibers, cold-drawing is also an efficient and convenient method; that is, drawing is performed under the condition that the material is at room temperature. Cold-drawn samples have the advantages of higher dimensional accuracy and better surface finish than hot-formed samples. As shown in Figure 6.4, during cold-drawing, the end of the sample is thinned so that it can pass through the wire drawing die hole. Then the hard slag shell on the surface is removed, and the lubricant is added, and finally, the target sample enters the wire drawing die hole for cold-drawing. The wire pulling speed should be controlled, and the continuous pulling should not exceed three times. If repulling is required, low-temperature annealing treatment should be used to eliminate internal stress.

Huo et al. used cold-drawing to process CoCrFeNi HEA (9.3 mm in diameter and about 16 µm in grain size) into wires with a diameter of about 7 mm [8], and the

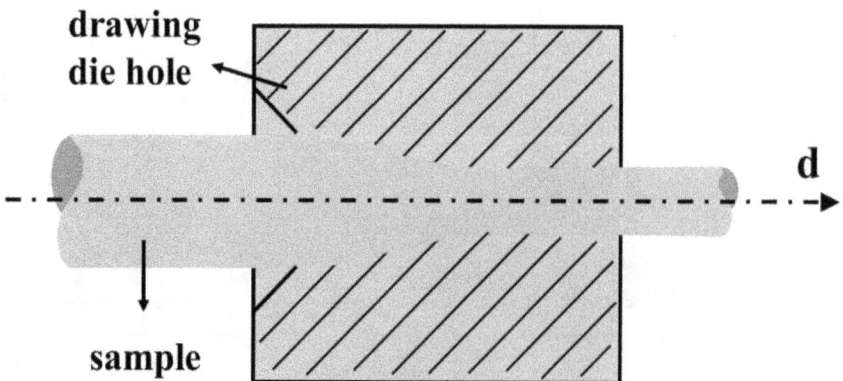

FIGURE 6.4 The working principle of cold-drawing.

strain rate was about $1 \times 10^{-1} - 1 \times 10^{1}$ s^{-1}. The HEA wire exhibits a large number of nano-twins. The authors characterize its mechanical behavior at 223–923 K and analyze its mechanism. The wire has the highest tensile yield strength (1.2 GPa) and elongation (13.6%) at 223 K. The yield strength and elongation decrease slightly at room temperature, and its strength remains at a relatively high level when the temperature is increased to 923 K, more than 800 MPa. Further transmission electron microscopy characterizations demonstrate that this is due to the hindering effect of primary and secondary nanoscale twin boundaries on dislocation slip. Also due to this effect, the alloy does not show strain hardening after yielding but softens.

Ma et al. performed cold-drawing and annealing of CoCrFeMnNi HEA wire in turn, and then cold-drawn with different strains, and studied its microstructure changes and related mechanisms in detail [9]. The wire diameter after the initial cold-drawing and annealing was 8 mm. The study found that the alloy wire corresponds to different deformation mechanisms under different strains. At low strains, in-plane slip is activated, and further slip bands are formed. At moderate strain, dislocation cells are formed, and the slip mode changes from plane slip to wave slip while the twinning density increases. When the amount of strain is further increased, a large number of deformation twins are formed, and at the same time, finer secondary twins are generated. Along with the evolution of the deformation mechanism, the wire yield strength increases and elongation decreases due to the proliferation of dislocations, the formation of Taylor lattices, and the contribution of twinning and texture.

Through a series of cold-drawing dies and multi-pass cold-drawing, Cho et al. drew the $Co_{10}Cr_{15}Fe_{25}Mn_{10}Ni_{30}V_{10}$ HEA wire with a diameter of 4.75 mm into a wire with a minimum diameter of 1.0 mm [10]. The drawing speed is 3 mm/s, and the cross-sectional reduction rate per pass is about 9.75%. Figure 6.5 shows photographs of continuous cold-drawn wire at different reductions of 0, 60, 80, 90, and 96%. The authors also studied the evolution of microstructure and mechanical properties under different deformations. At the same time, they characterized the cross-sectional microstructure and texture of different strained wires and further revealed the differences in the core and surface microstructures. The stress and strain along the radial direction of the surface and the core of the wire are different, resulting in different structures. With the increase of deformation, <111> and <100> textures mainly appear in the core structures, while random textures appear on the surface. Figures 5(f) and (g) show that the hardness at the core and surface increases with the deformation, and the hardness at the surface is always higher than that at the center. This hardness distribution may be due to the larger deformation of the surface during the cold-drawing process, resulting in grain refinement and the appearance of more nano-deformation twins.

Kwon et al. fabricated CoCrFeMnNi equiatomic HEA fibers at 77 K by a cryogenic drawing process [11], using 11 circular holes with progressively smaller diameters to reduce the diameter of the rod from 12.5 mm to 7.5 mm, and the total area was reduced by 64%. The HEA fiber contains a large number of nano-twins, including primary twins and secondary twins, which further refines the grains and obtains a substantial increase in strength. Compared with the initial ingot before drawing, the fibrous yield strength after drawing is increased by 4.7 times, from 0.328 GPa to 1.54 GPa, and the tensile strength is increased 2.3 times. The current mechanical

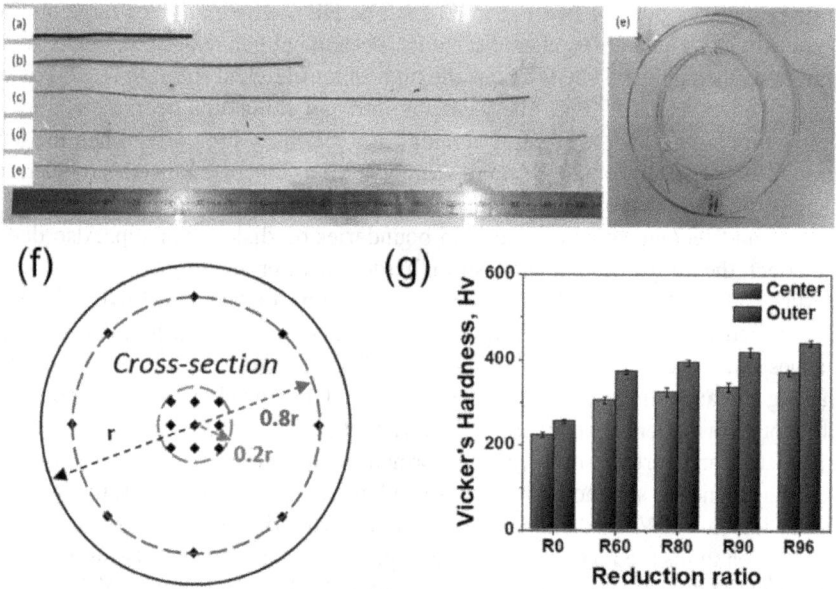

FIGURE 6.5 (a)–(e) Macroscopic views of cold-drawn fibers with diameter from 1 mm to 4.75 mm. (f) Positions of hardness testing at outer and center. (g) Hardness values of the outer and center of the fiber [10].

properties are better than the current commonly used pearlitic steel and tempered martensitic steel. Then the fiber was filled with different contents of hydrogen to test its resistance to hydrogen embrittlement, and it was found that CoCrFeMnNi HEA has better hydrogen embrittlement resistance performance than pearlitic steel.

6.4 HEA FIBERS PREPARED BY THE TAYLOR-ULITOVSKY METHOD

Taylor first invented the preparation method of glass-coated metallic filaments in 1924 [12], so the glass-coated melt spinning method is also called the Taylor method. Because it is difficult to accurately control the temperature of the metallic molten pool and the glass tube in the early Taylor method, the continuous length of the prepared filaments can only reach a few meters or even only a few centimeters. In the 1950s and 1960s, Ulitovsky made a relatively large improvement on the Taylor method, which could achieve semi-automatic production after improvement, so the glass-coated melt spinning method is also called the Taylor-Ulitovsky method.

The basic principle of the Taylor-Ulitovsky method is shown in Figure 6.6. The metal is first put into a glass tube, evacuated, and then filled with argon. Subsequently, the glass tube enters the induction heating zone at a certain speed, and the high-frequency induction heater is used to melt the metal and soften the end of the glass tube. The end of the softened glass tube is drawn into a very thin capillary tube, and the molten metal enters the capillary tube, and after rapid solidification, glass-coated

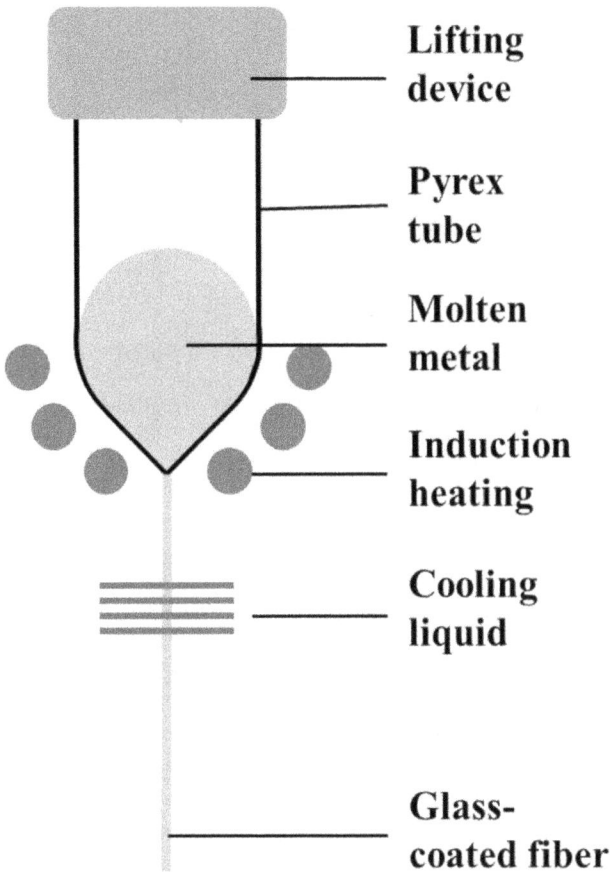

FIGURE 6.6   The working principle of the Taylor-Ulitovsky method.

metallic filaments are obtained. This technology utilizes the characteristics of glass with high viscosity, low surface tension, and easy drawing and forming. At the same time, softening the glass can protect the molten metal from oxidation.

Chen et al. successfully prepared CoCrNi fibers with diameters of 100 and 40 μm, respectively, by using the Taylor-Ulitovsky technique [13]. The fibers prepared by the Taylor-Ulitovsky method are all composed of randomly oriented equiaxed crystals (with a small number of twins), and the average grain size decreases from 7.5 μm to 5.1 μm as the diameter decreases. In terms of mechanical properties, compared with the as-cast bulk samples, the fibers have greatly improved yield strength and tensile strength, and only a small number sacrificed ductility. As the diameter decreases, the tensile strength and ductility are further improved. When the diameter of the fibers decreased from 100 μm to 40 μm, the yield strength increased from 450 MPa to 638 MPa, the ultimate tensile strength increased from 950 MPa to 1,188 MPa, and the tensile ductility also increased from 41% to 48%. Similar to other HEA fibers, the previously mentioned excellent mechanical properties originate from dislocation

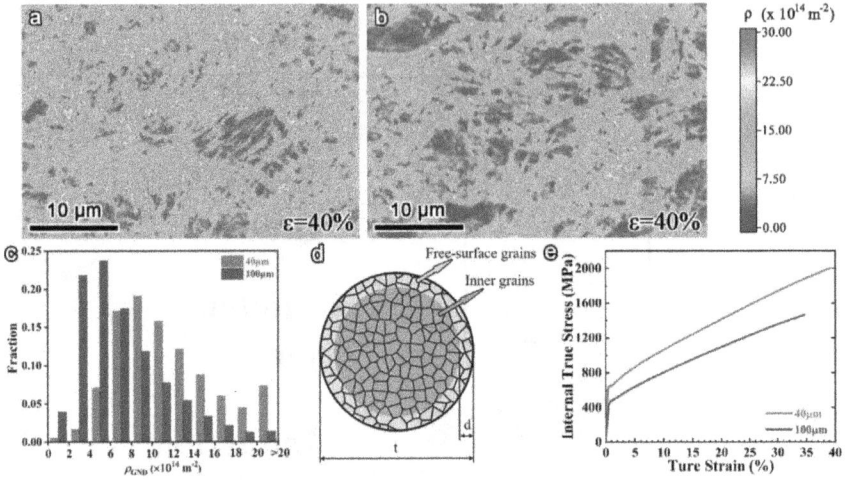

FIGURE 6.7 The GND densities of (a) 40-microwire and (b) 100-microwire with strain of 40% based on the KAM; (c) The distribution of GND density; (d) Schematic of microwires within cross-section; (e) True stress-strain curves of inner grains. Here, the icon on the right indicates the smallest density in blue and the largest density in red, and the density increases from blue to red [13].

interactions, deformed nano-twins, martensitic transformation, HCP stacking faults, and others.

The CoCrNi alloy fibers have a large size effect; that is, the strength and plasticity of the 40 μm diameter fiber are significantly improved compared with the 100 μm diameter fiber. The authors analyzed it using the KAM plot. As shown in Figure 6.7, under the same 40% strain condition, the geometrically necessary dislocation density in the 100 μm diameter fiber is 7.1×10^{14} m^{-2}, while that in the 40 μm fiber is as high as 12.3×10^{14} m^{-2}. The higher density of geometrically necessary dislocations (GNDs) leads to higher strain gradients, which in turn are linked to multiple deformation twins, resulting in high strength and ductility in the 40 μm fibers. At the same time, the dislocation distribution in the 40 μm fiber is more uniform, which limits the local deformation and leads to uniform plastic flow.

6.5 SUMMARY AND PROSPECT

This chapter summarizes the preparation methods and mechanical properties of the currently reported representative HEA fibers and analyzes their strengthening mechanisms. At present, HEA fibers are mainly prepared by hot-drawing, cold-drawing, and Taylor-Ulitovsky glass-coating method, and mainly focus on the face-centered cubic HEA system dominated by CoCrNi. Depending on the melting point of the selected HEA, hot-drawing is usually performed at around 1073 K, and recrystallization occurs during the drawing process. The diameters from

1 mm to 10 mm have been reported. Cold-drawing is carried out at room temperature. The cold-drawing process accumulates a large amount of plastic deformation. The extremely high dislocation density makes the prepared wire have high strength and poor plasticity, but the mechanical properties can be improved by annealing regulation. The diameter of the reported cold-drawn fiber is between 1 mm and 10 mm. The Taylor-Ulitovsky method is an effective method for preparing micron-scale fiber, where the diameter can reach 40 microns, and it can be formed from molten metal at one time, with the advantages of a short process and convenient operation.

The HEA fibers prepared by these methods have excellent mechanical properties at room temperature and low temperature, surpassing other fibers, such as existing high-strength steel wires, and have the advantage of being stronger and tougher at lower temperatures. Their high strengths are mainly due to the joint action of dislocations and twins, and also include the effects of solid-solution strengthening, grain-refinement strengthening, nanoprecipitation strengthening, and phase-transformation strengthening.

Based on its excellent room-temperature and low-temperature strength-plasticity matching, HEA fibers have promising applications in power electronics, aerospace, and other industries. At the same time, diversified preparation methods of different systems of HEAs can be further developed, and their electrical, magnetic, and other functional properties can be studied. In addition, HEA fibers can also be applied to fiber-reinforced composite materials, which can enhance and toughen the matrix and help reduce costs.

REFERENCES

[1] Zhang W, Liaw P K, Zhang Y. Science and technology in high-entropy alloys[J]. *Science China Materials*, 2018, 61(1): 2–22.

[2] Lu C, Niu L, Chen N, et al. Enhancing radiation tolerance by controlling defect mobility and migration pathways in multicomponent single-phase alloys[J]. *Nature Communication*, 2016, 7: 13564.

[3] Luo H, Lu W, Fang X, et al. Beating hydrogen with its own weapon: Nano-twin gradients enhance embrittlement resistance of a high-entropy alloy[J]. *Materials Today*, 2018, 21(10): 1003–1009.

[4] Chen J, Zhou X, Wang W, et al. A review on fundamental of high entropy alloys with promising high—temperature properties[J]. *Journal of Alloys and Compounds*, 2018, 760: 15–30.

[5] Li D, Li C, Feng T, et al. High-entropy $Al_{0.3}CoCrFeNi$ alloy fibers with high tensile strength and ductility at ambient and cryogenic temperatures[J]. *Acta Materialia*, 2017, 123: 285–294.

[6] Li D, Gao M C, Hawk J A, et al. Annealing effect for the $Al_{0.3}CoCrFeNi$ high-entropy alloy fibers[J]. *Journal of Alloys and Compounds*, 2019, 778: 23–29.

[7] Liu J-P, Chen J-X, Liu T-W, et al. Superior strength-ductility CoCrNi medium-entropy alloy wire[J]. *Scripta Materialia*, 2020, 181: 19–24.

[8] Huo W, Fang F, Zhou H, et al. Remarkable strength of CoCrFeNi high-entropy alloy wires at cryogenic and elevated temperatures[J]. *Scripta Materialia*, 2017, 141: 125–128.

[9]   Ma X, Chen J, Wang X, et al. Microstructure and mechanical properties of cold drawing CoCrFeMnNi high entropy alloy[J]. *Journal of Alloys and Compounds*, 2019, 795: 45–53.
[10]   Cho H S, Bae S J, Na Y S, et al. Influence of reduction ratio on the microstructural evolution and subsequent mechanical properties of cold-drawn Co10Cr15Fe25Mn10Ni30V10 high entropy alloy wires[J]. *Journal of Alloys and Compounds*, 2020, 821.
[11]   Kwon Y J, Won J W, Park S H, et al. Ultrahigh-strength CoCrFeMnNi high-entropy alloy wire rod with excellent resistance to hydrogen embrittlement[J]. *Materials Science and Engineering: A*, 2018, 732: 105–111.
[12]   Taylor G F. A method of drawing metallic filaments and a discussion of their properties and uses[J]. *Physical Review*, 1924, 23(5): 655–660.
[13]   Chen J-X, Chen Y, Liu J-P, et al. Anomalous size effect in micron-scale CoCrNi medium-entropy alloy wire[J]. *Scripta Materialia*, 2021, 199: 113897.

7 High-Entropy Powder and Hard-Cemented Carbide Alloys

Yong Zhang and Yuxin Wen

7.1 INTRODUCTION

As a new functional material, the high-entropy alloy shows excellent comprehensive properties. Because of its unique structure, the high-entropy alloy can inhibit the growth of hard particles and produce the effect of fine-grain strengthening. Its high stability ensures the stability of the bonding phase and hard phase in the sintering process, and there will not be a lot of harmful phases that reduce the performance of cemented carbide. High-entropy alloy can also meet the needs of different working conditions through the coordination of different elements, is a potential excellent bonding material, and can achieve the purpose of reducing Co or replacing Co so as to reduce the production cost of hard alloy, improve the performance of hard alloy, and promote the wide application of hard alloy. Professor Zhang Yong's team used GS201 ($AlCoCrFeNiTi_{0.2}$) and GS301 ($AlCo_{0.4}CrFeNi_{2.7}$) as the bonding phase, and WC as the hard phase to prepare high-entropy cemented carbides, and studied the properties of these alloys.

7.2 POWDER PROCESSING

High-entropy alloys have been found to have excellent mechanical properties under some cold-rolling, annealing, and other processes. At present, the preparation methods of high-entropy alloys are more focused on the preparation of bulk materials, and the bulk materials of high-entropy alloys are mostly composed of single-phase BCC, FCC, or HCP crystal structures. If you want to obtain a multiphase high-entropy alloy, it is necessary to prepare a uniform high-entropy alloy powder. Powder metallurgy has been widely used in the preparation of high-entropy alloys due to its advantages of net forming, low cost, and high efficiency, and the first step of the powder metallurgy process is to obtain the powder suitable for the process. Many processing methods for high-entropy alloy powder have been developed. This chapter summarizes several of the most commonly used processing methods of high-entropy alloy powder. It mainly includes the ball mill mechanical alloying method and aerosol method (including EIGA and VIGA).

DOI: 10.1201/9781003319986-7

7.2.1 MECHANICAL ALLOYING OF BALL MILL

Mechanical alloying by ball milling is the most commonly used technique for obtaining high-entropy alloy powders. In this technology, the powder particles are repeatedly cold-welded, broken, and rewelded through high-energy ball milling so that each element is alloyed at the atomic level to obtain the solid powder of the alloy. The advantages of ball mill mechanical alloying technology are its simplicity and low cost, and the high-entropy alloy powder obtained by this technology has a stable microstructure, excellent chemical homogeneity, and room-temperature machining performance.

The ball mill mechanical alloying method is also known as the solid-state and non-equilibrium manufacturing method. At present, CoCrFeMnNi [1], AlCrCoFeNi [2], AlCoCuZnNiTi [3], AlCrCuFeTiZn [4], AlFeCuCrMg$_x$ (x = 0, 0.5, 1), AlCrCuFeTiZn [4], and AlFeCuCrMg$_x$ have been successfully prepared by the mechanical alloying method (1.7 mol) [5] high-entropy alloy powder. In 2007, Indian scholar S.Vanalakshmi [4] prepared AlFeTiCrZnCu high-entropy alloy by the mechanical alloying method for the first time. Y.-L. Chen et al. [6] prepared a Cu$_{0.5}$NiAlCoCrFeTiMo high-entropy alloy by the mechanical alloying method and studied the competition of components in this system in mechanical alloying. It is found that mechanical alloying can not only prepare FCC or BCC phase high-entropy alloy powder with a single-phase crystal structure but also prepare FCC and BCC biphase high-entropy alloy powder. However, this method has some disadvantages, such as a long ball milling time, poor sphericity of particles, and easy contamination of the powder.

Zhou et al. prepared a CoCrFeNiTiCuMo$_x$V$_x$ high-entropy alloy powder with BCC and FCC biphase structure by mechanical alloying with a ball mill and found that when x = 0.5, 1.0, the diffraction peaks of BCC (110),(211) crystal plane and FCC (111) crystal plane appear obviously in the powder, but when x = 1.5, 2.0, the diffraction peaks of BCC (200) crystal plane also appear, and with the increase of Mo and V, BCC becomes the main phase [7].

The ball milling time of mechanical alloying can also affect the structure, shape, and grain size of the high-entropy alloy powder. The influence of ball milling time on the high-entropy NbMoTaW alloy powder was studied by Qipeibu et al. [8]. It can be seen from Figure 7.1 that with the extension of the ball milling time, the grain size decreases and the lattice distortion increases and, eventually, tends to a stable value. The morphology of the powder was characterized at different milling times. It was found that the powder had good sphericity and uniform particle size at 45 h. Figure 7.2 shows the SEM image of the high-entropy alloy powder after ball milling.

7.2.2 AEROSOL METHOD (EIGA AND VIGA)

In order to obtain a powder with better sphericity, the high-entropy alloy powder can be prepared by the aerosol method. The gas atomization method uses gas as an atomizing medium to impact the molten metal or alloy liquid flow to break it into fine droplets and then quickly solidify it to produce a powder.

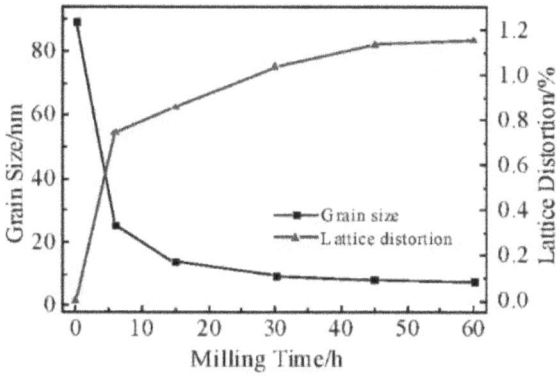

FIGURE 7.1 Relationship of grain size and lattice distortion rate with ball milling time [8].

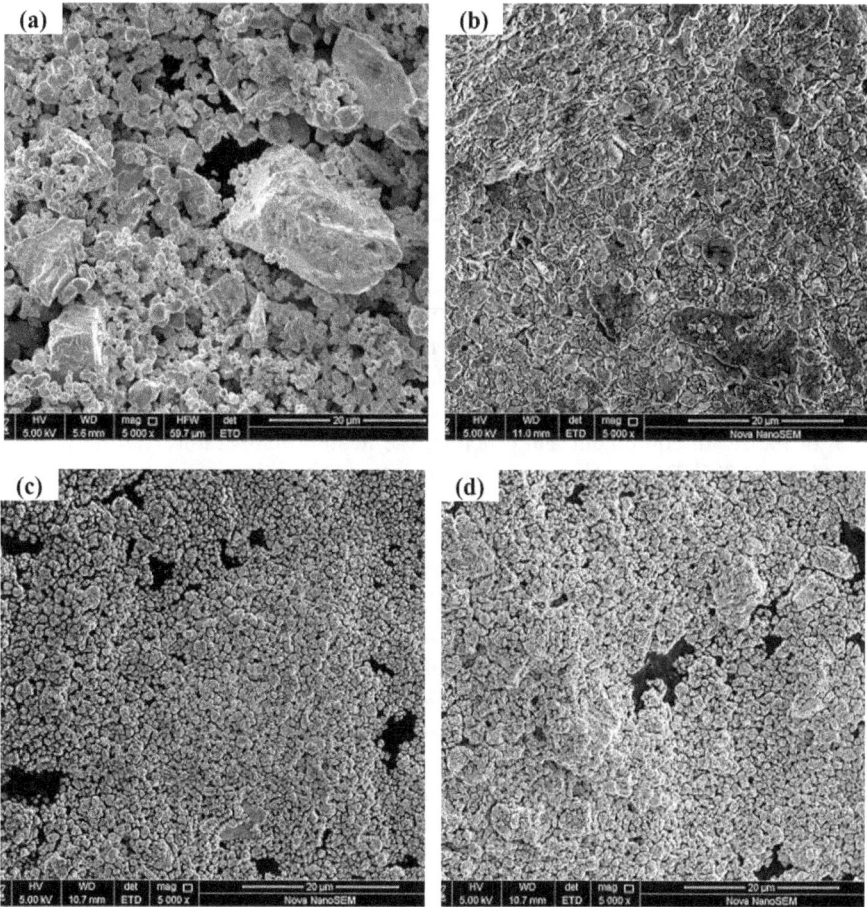

FIGURE 7.2 SEM morphologies of NbMoTaW HEA powder at different milling times: (a) 0 h, (b) 6 h, (c) 15 h, (d) 30 h, (e) 45 h, and (f) 60 h.

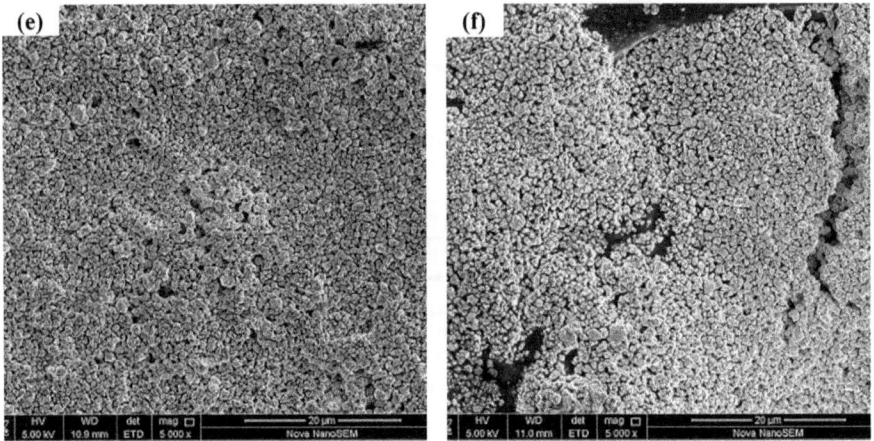

FIGURE 7.2 (Continued)

Aerosol methods include EIGA and VIGA. EIGA is a process in which a prefabricated alloy rod is used as an electrode, and the rotating rod electrode is melted and atomized by the induction of a melting coil and parameters that control the vertical feeding rate. This method avoids the non-metallic impurities in the traditional crucible melting process and, thus, significantly improves the purity of the atomized powder. Figure 7.3 shows the atomization principle of EIGA.

The alloy liquid flows through the tube at the bottom of the furnace to the atomizing nozzle, where it is crushed by supersonic gas impact and atomized into micrometer-scale fine droplets, which spheroidize and solidify into powder. Figure 7.4 is the schematic diagram of VIGA powder making. Due to the limitation of hardware equipment and crucible, the heating temperature can only reach 1500°C~1600°C. Due to the influence of the ceramic crucible and nozzle, impurities will be substituted into the alloy melt, which will affect the purity of the prepared metal powder.

Peipei Ding et al. prepared an $AlCoCrFeNi_{2.1}$ high-entropy alloy powder by gas atomization method and found that the $AlCoCrFeNi_{2.1}$ high-entropy alloy powder obtained by atomization method had a different crystal structure from the bulk high-entropy alloy [9]. According to XRD analysis, as shown in Figure 7.5, the powder structure of $AlCoCrFeNi_{2.1}$ is a biphasic structure of FCC and BCC, and then the SEM/BSE image of the atomized powder section is analyzed, as shown in Figure 7.6. It can be seen from Figure 7.6(a) that the particles of $AlCoCrFeNi_{2.1}$ are composed of a massive phase and a dendrite phase. It can be seen from Figure 7.6(b) that the particles have a biphase structure, which also confirms the conclusion of XRD. Lu J et al. also prepared an AlCoCrFeNi high-entropy alloy powder by the aerosol method and found that the powder prepared by this method had a very low oxidation rate and very high oxidation activation energy. Mr. Zhang Yong's team prepared high-entropy alloy powders of GS102, GS2011($AlCoCrFeNiTi_{0.2}$), and GS301($AlCo_{0.4}CrFeNi_{2.7}$) by

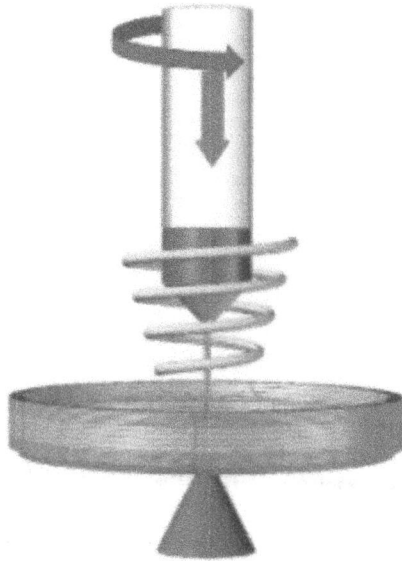

FIGURE 7.3   Schematic diagram of the EIGA atomization principle.

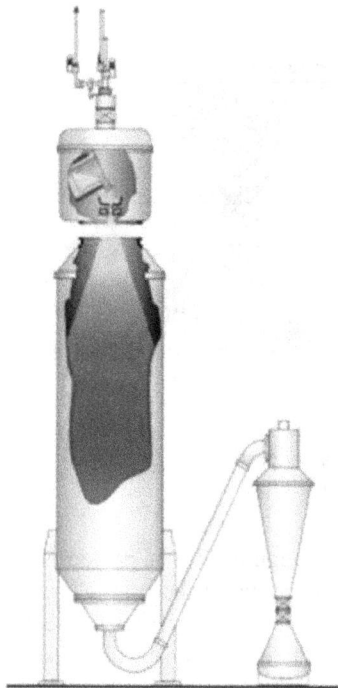

FIGURE 7.4   Schematic diagram of VIGA powder making.

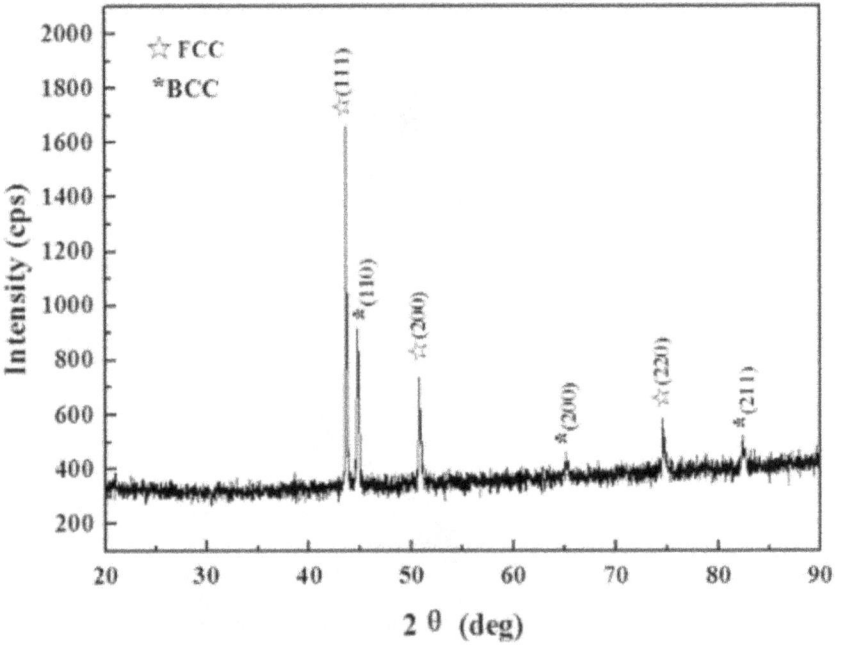

FIGURE 7.5   XRD patterns of AlCoCrFeNi$_{2.1}$ high-entropy alloy powder [9].

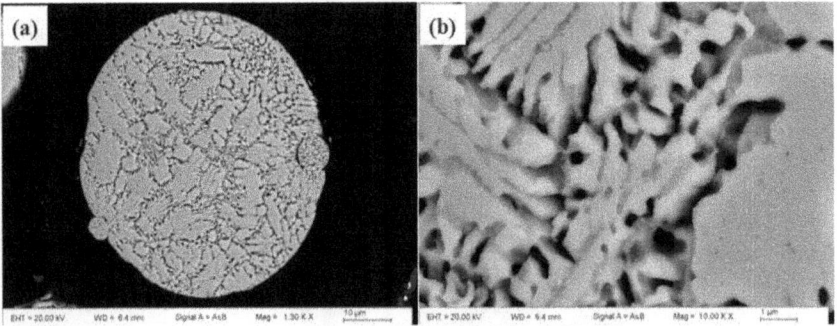

FIGURE 7.6   Backscattered SEM images of the cross-section of the as-atomized AlCoCrFeNi$_{2.1}$ high-entropy alloy particle: (a) the crystalline structure in a particle and (b) the lamellar dual-phase structure [9].

aerosol method, as shown in Figure 7.7. The powder surfaces of GS201 and GS301 are smooth and round. However, the surface of the large-size powder has a dendritic morphology. The surface of the small particles of GS102 is smooth, the surface of large particles is rough, and there is no obvious characteristic morphology.

In the past, high-entropy alloys were mainly made by mechanical alloying and casting, but the high-entropy alloys made by traditional methods have many defects,

FIGURE 7.7　High-entropy alloy powder morphology: (a) GS102, (b) GS201, and (c) GS301.

such as high porosity. Nowadays, additive manufacturing is becoming more and more popular in the field of metal processing, which can avoid the defects of pores brought by the traditional method. The performance of the alloy prepared by this method is often closely related to the quality of the metal powder. The powder prepared by the aerosol method has the advantages of high purity, low oxygen content, and fewer impurities, so it is often used as the preferred material for additive manufacturing.

Yang Chichao et al. [10] prepared AlCoCrCuFeNiSi high-entropy alloy powder by aerosol method (Figure 7.8 and Figure 7.9). According to SEM microscopic images, the powders prepared by the aerosol method are spherical with smooth surfaces. As can be seen from the XRD pattern, the powder obtained by the aerosolization method shows a sharper diffraction peak and a reduced full width at the half peak, indicating that it has higher crystallinity than the casting alloy. This is because the aerosolized alloy powder formed a single-phase solid solution at a faster solidification rate.

At present, whether the high-entropy alloy powder is prepared by mechanical alloying or aerosolization, subsequent consolidation and other treatments are needed for it to have an application value. The performance of the high-entropy alloy prepared by these two methods is much better than that of the traditional melting method. However, as a new material, there are many limitations in the application of a high-entropy alloy. It is believed that with the solution to the problem, high-entropy alloys will soon have value in real-world applications.

7.3　PROCESSING OF HIGH-ENTROPY CEMENTED CARBIDES GS201 AND GS301

Due to the similar properties of Co, Cr, Fe, and Ni, the four metals are easy to be formed into a solid-solution structure, so the high-entropy alloy containing these four elements is the most widely studied, and the study shows that the addition of an Al element can make the phase structure gradually change from FCC to BCC, the microstructure of the alloy tends to be simplified, and the hardness and wear resistance of the alloy are enhanced. Due to the BCC structure of the AlCoCrFeNi alloy, it has high strength and hardness, but its plasticity is limited. The biphase BCC+FCC alloy system has both high strength and good plasticity. Therefore, Professor Zhang Yong's research group chose AlCoCrFeNi for further research. By adding trace elements

(a)

(b)

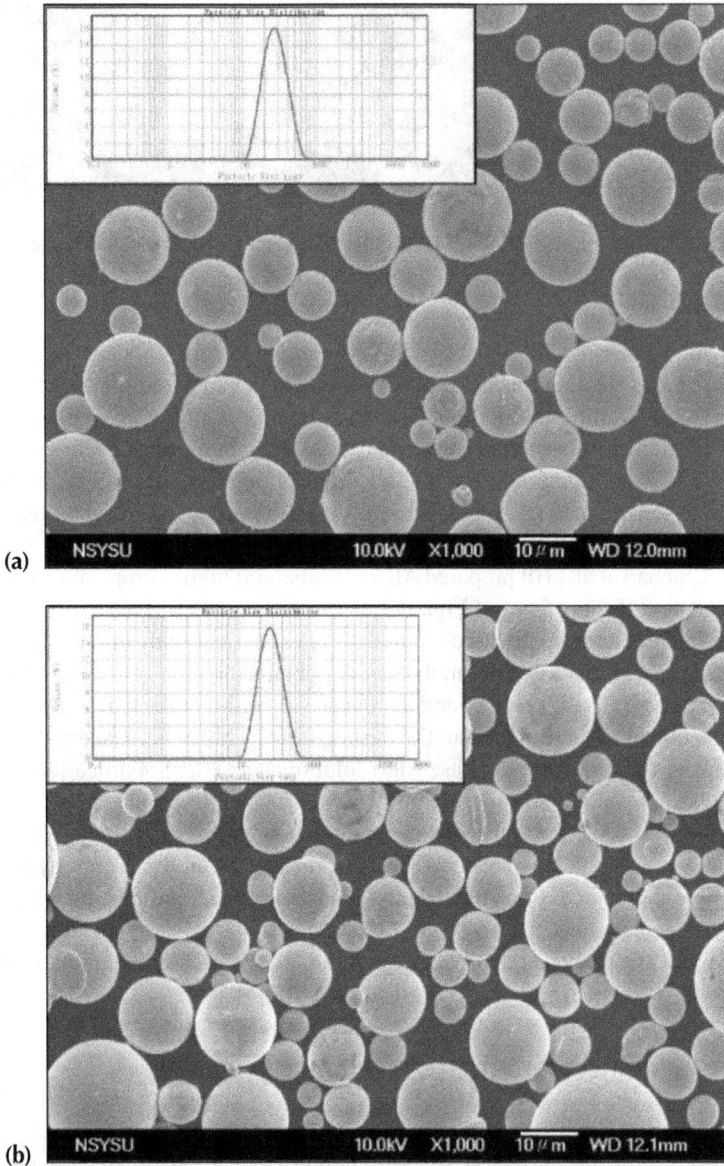

FIGURE 7.8 SEM micrograph and particle size distribution (inset) of (a) $Al_{0.5}CoCrCuFeNi$ alloy powders, (b) $Al_{0.5}CoCrCuFeNiSi_{1.2}$ alloy powders, and (c) $Al_{0.5}CoCrCuFeNiSi_{2.0}$ alloy powders [10].

and adjusting the composition to achieve entropy control, the excellent performance of GS201 ($AlCoCrFeNiTi_{0.2}$) and GS301 ($AlCo_{0.4}CrFeNi_{2.7}$) alloys was developed. The traditional cemented carbide always has the disadvantage of high hardness and low toughness.

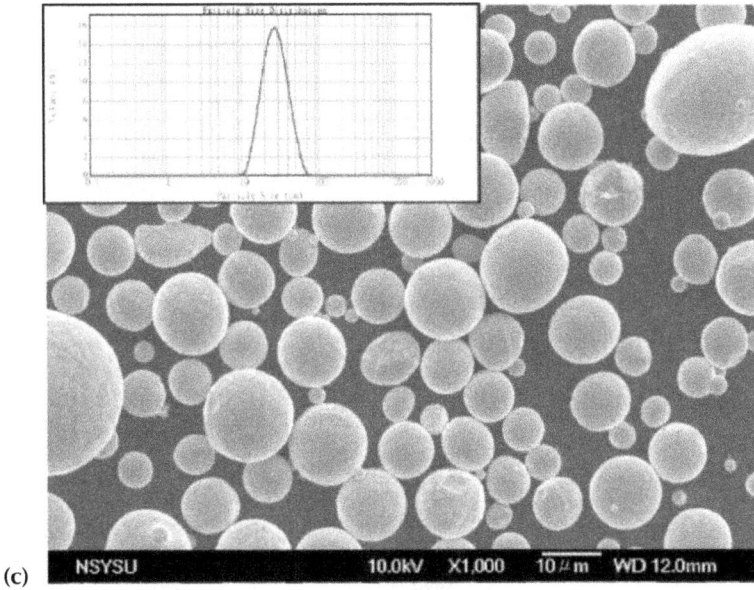

(c)

FIGURE 7.8   (Continued)

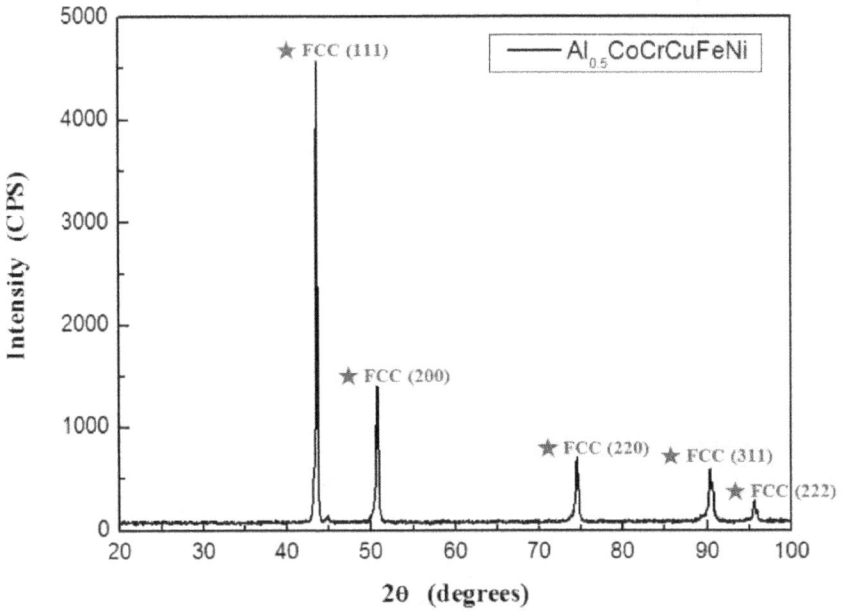

FIGURE 7.9   XRD patterns of $Al_{0.5}CoCrCuFeNi$ alloy powders [10].

Recent studies have shown that as a new functional material, the high-entropy alloy has excellent properties in strength, hardness, wear resistance, and low-temperature toughness due to its unique high-entropy effect, slow diffusion effect, lattice distortion effect, and cocktail effect. Using it as a binder can effectively solve the problem of hard alloys being both hard and tough.

7.3.1 Processing of High-Entropy Alloys GS201 and GS301

The high-entropy alloy GS201 is based on AlCoCrFeNi alloy by adding a trace amount of Ti element to increase the entropy of the alloy. Moreover, due to the large atomic radius of the Ti element, the strength and hardness of the alloy increased, and the plasticity also increased. Professor Zhang Yong's research group prepared $AlCoCrFeNiTi_x$ ($x = 0, 0.1, 0.2, 0.3, 0.4, 0.5$) alloy to study the influence of the change of entropy on the microstructure through room-temperature compression test and hardness test and to study the influence of entropy regulation on the mechanical properties of the alloy. GS201 ($AlCoCrFeNiTi_{0.2}$) is selected as the alloy component with the best performance. When $x = 0.2$, the maximum compressive plasticity of the alloy is 32.59%, the compressive yield strength is 1530.42 MPa, the compressive strength is 4035.00 MPa, and the hardness is 597.8 HV. The vacuum arc melting method is used for the preparation of GS201 alloy. The electronic weighing of the required mass of each element with an accuracy of 0.001 g is used. Each component is put into aqueous ethanol and cleaned in the ultrasonic cleaning instrument, and then each element is put into the crucible of the vacuum arc furnace in turn. Close the furnace door, open the mechanical pump to pump vacuum when pumping below 5 Pa, and open the molecular pump to pump vacuum below 104 Pa. When the speed of the molecular pump reaches 600 r/min, shut down the molecular pump, fill it with high-purity argon, and melt the titanium block by arc. If the titanium block does not change color, it indicates that the vacuum degree has met the smelting requirements. At this time, the required alloy is melted, and the alloy is melted 4 to 5 times to ensure a uniform structure.

Professor Zhang Yong's research group studied the GS301 $AlCo_xCrFeNi_{3.1-x}$ ($x = 0.4, 1$) alloy system. Under the condition of keeping Al, Cr, and Fe unchanged, GS301 increases the content of Ni and reduces the content of Co to regulate entropy. By studying its structure and mechanical properties, it is found that $AlCo_{0.4}CrFeNi_{2.7}$ has excellent comprehensive performance. GS301 is made by vacuum induction melting and has high-purity elements (Al, Co, Ni: 99.9 wt.%; Cr, Fe: 99.5–99.6 wt.%). All the ingredients are heated to 600°C in a ZrO_2 crucible and kept for 1 hour to remove water vapor. The pouring temperature was set to 1500°C, and the IRTM-2CK infrared pyrometer was used to monitor the temperature with an absolute accuracy of ±2°C. About 2.5 kilograms of metal was melted, superheated, and poured into a crucible of high-purity graphite with a length of 220 mm, an upper inner diameter of 62 mm, and a bottom inner diameter of 50 mm. In all cases, the furnace chamber was first evacuated to 6×10^{-2} Pa and then backfilled with high-purity argon up to 0.06 MPa.

7.3.2 Processing of High-Entropy Cemented Carbides GS201 and GS301

Hard alloy is a powder metallurgy product sintered in a vacuum furnace or hydrogen reduction furnace with high hardness refractory metal carbide (WC, TiC) micron powder as the main component and Co, Ni, Mo as a binder.

Nearly 65% of total cemented carbide production is related to metal cutting tools. The mining, oil drilling, and rock industries account for about 15% of the market, while the timber and construction industries account for 10%. Compared with other hard materials, cemented carbide accounts for 50% of the world's total market, and its further widespread application requires reducing its production cost. The wear resistance of the traditional hard alloy is poor, and the hardness of a hard alloy will decrease due to the existence of the adhesive phase at a high temperature. The price of cobalt, a resource-poor metal used as a traditional bond, has also risen with its use in electric vehicles as a raw material for batteries. In addition to the performance of traditional cemented carbide not meeting people's production needs for cobalt as a binder, the production cost of cemented carbide is gradually increased, coupled with cobalt carcinogenicity, it is urgent to find new materials to replace cobalt in the application of cemented carbide. In recent years, with the development of the field of high-entropy alloy, it has shown excellent comprehensive properties and is considered a potential cemented carbide phase.

Chen et al. [11] prepared a WC-based cemented carbide with $Al_{0.5}CoCrCuFeNi$ as the bonding phase. It is found that the hardness of WC/$Al_{0.5}CoCrCuFeNi$ high-entropy alloy is higher than that of W because the $Al_{0.5}CoCrCuFeNi$ high-entropy alloy can significantly inhibit the growth of WC grains as a bonding phase, and the spacing of WC grains increases with the increase of the content ratio of $Al_{0.5}CoCrCuFeNi$ alloy C/Co traditional materials (Figure 7.10)

Zhou et al. studied the influence of AlFeCoNiCrTi high-entropy alloy as a bond relative to ultrafine-crystal WC-based cemented carbides [12]. The study also shows that the bonding of high-entropy alloy can inhibit the growth of WC grains, thus increasing its hardness and corrosion resistance. In addition, the Vickers hardness of WC-20 AlFeCoNiCrTi alloy studied by Zhou et al. is twice that of WC-20 $Al_{0.5}CoCrCuFeNi$ alloy studied by Chen et al. Fu et al. [13] used TiNiFeCrCoAl as the bonding phase to prepare TiB_2-based cemented carbides. The study showed that HEA reacted with the oxide in TiB_2 at a high temperature, thus improving the density of the alloy. The hardness of cemented carbide decreases with the increase of HEA content, but the fracture toughness is the opposite. Obra et al. [14] prepared Ti (C, N)-based cemented carbides using CoFeNi high-entropy alloy as the bonding phase. The results showed that the volatilization of Mn and a high degree of segregation of Cu would lead to the increase of the porosity of the alloy, thus deteriorating the properties. The presence of V will promote the generation of characteristic nuclear edge microstructure of Ti (C, N) foundation ceramics.

Among the high-entropy alloys, the AlCoFeNi-X series high-entropy alloy is the most studied system. Zhang Yong's research group chose AlCoFeNi-X series high-entropy alloy as the bonding phase to prepare cemented carbide. WC-GS201 and WC-GS301 composites with different sintering temperatures and binder content were prepared by discharge plasma sintering using GS201 and GS301 as binder phase,

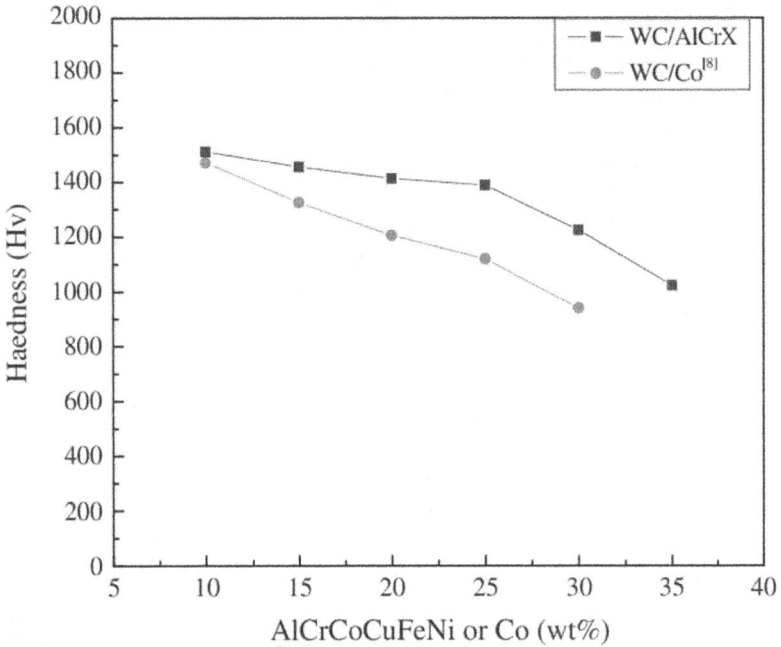

FIGURE 7.10 The hardness change curves of different ratios of WC and binder metal composites [11].

respectively. It is found that when the binder phase content is 5–10%, the gap of WC can be fully filled. When GS201 is used as the binder phase, the high-entropy cemented carbide sintered at 1300°C has the best mechanical properties, with a hardness of 1567.8 HV_{30} and fracture toughness of 7.5 MPa m$^{1/2}$. When GS301 is used as the bonding phase, the high-entropy cemented carbide sintered at 1300°C has the best mechanical properties, and the hardness and fracture toughness are 1575 HV_{30} and 9.2 MPa m$^{1/2}$, respectively. In addition, although the hardness of GS201 is higher than that of GS301, under the same conditions, as the cemented carbide phase, GS201 does not show this advantage in hardness. On the contrary, GS301 has higher hardness and fracture toughness due to the formation of relatively higher density blocks.

It can be seen from Figure 7.11 that the hardness of WC-5% GS301 alloy decreases with the increase of sintering temperature, while the fracture toughness is on the contrary. Further, the relationship between the hardness and fracture toughness of WC-HEA and the bonding phase content is studied at 1300°C. It can be seen from Figure 7.12 that the Vickers hardness of the alloy decreases with the increase of bonding phase content. At the same time, Zhang Yong's research group also proposed that the properties of high-entropy alloy are closely related to the relative density: the higher the relative density of the alloy, the fewer the defects caused by cracks, which is conducive to the improvement of density and toughness.

From the very beginning, the high-entropy alloy showed a series of excellent properties, such as high hardness and good toughness, which attracted the attention of researchers.

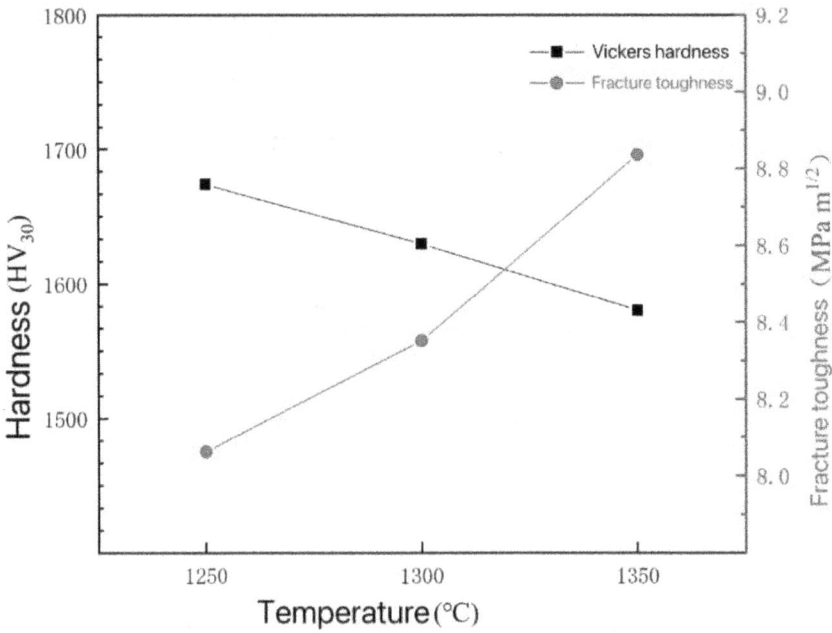

FIGURE 7.11   Relationship between hardness and fracture toughness of WC-5% GS301 and sintering temperature.

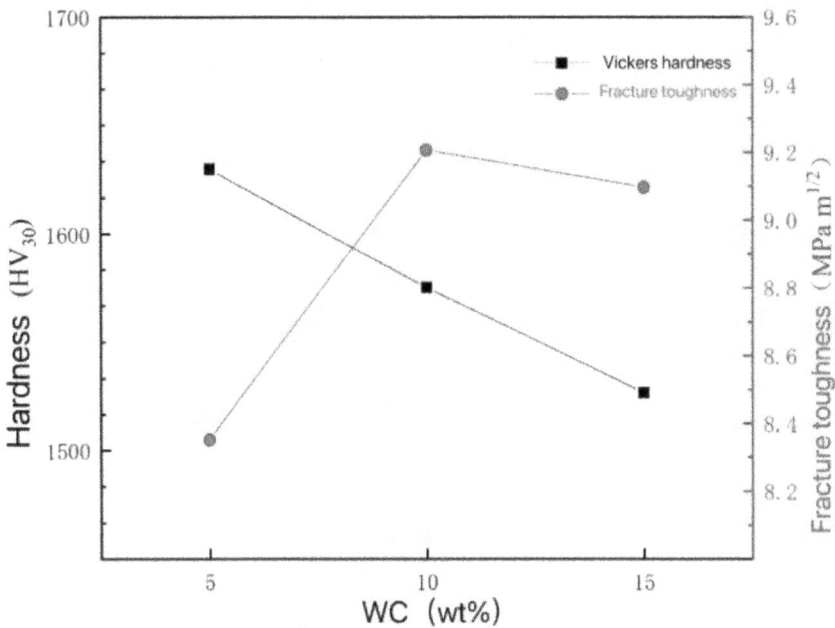

FIGURE 7.12   Relationship between hardness and fracture toughness of WC-HEA sintered at 1300°C and the content of bond phase.

Many studies have confirmed the possibility of the high-entropy alloy as a bonding phase, but some problems should be paid attention to in the selection of a high-entropy alloy, such as whether the high-entropy alloy as a bonding phase has good wettability with the hard phase matrix. The research of high-entropy alloys as cemented carbide is still in the initial state. The precipitation process of each component during sintering and the key issues on the microstructure and properties of the alloy remain to be studied in order to promote the application of high-entropy alloys in cemented carbide.

REFERENCES

[1]   Joo S H, Kato H, Jang M J, et al. Structure and properties of ultrafine-grained CoCrFeMnNi high-entropy alloys produced by mechanical alloying and spark plasma sintering[J]. *Journal of Alloys and Compounds*, 2017, 698: 591–604.

[2]   Ji W, Fu Z, Wang W, et al. Mechanical alloying synthesis and spark plasma sintering consolidation of CoCrFeNiAl high-entropy alloy[J]. *Journal of Alloys and Compounds*, 2014, 589: 61–66.

[3]   Varalakshmi S, Kamaraj M, Murty B S. Processing and properties of nanocrystalline CuNiCoZnAlTi high entropy alloys by mechanical alloying[J]. *Materials Science and Engineering a-Structural Materials Properties Microstructure and Processing*, 2010, 527(4–5): 1027–1030.

[4]   Varalakshmi S, Kamaraj M, Murty B S. Synthesis and characterization of nanocrystalline AlFeTiCrZnCu high entropy solid solution by mechanical alloying[J]. *Journal of Alloys and Compounds*, 2008, 460(1–2): 253–257.

[5]   Maulik O, Kumar D, Kumar S, et al. Structural evolution of spark plasma sintered AlFeCuCrMg$_x$ (x=0, 0.5, 1, 1.7) high entropy alloys[J]. *Intermetallics*, 2016, 77: 46–56.

[6]   Chen Y L, Hu Y H, Hsieh C A, et al. Competition between elements during mechanical alloying in an octonary multi-principal-element alloy system[J]. *Journal of Alloys and Compounds*, 2009, 481(1–2): 768–775.

[7]   Zhou J, Fan X F, Chen Y, Gou Y, Li Y Q. Microstructure of CoCrFeNiTiCuMo$_x$V$_x$ high entropy alloy Powder prepared by mechanical alloying[J]. *Materials for Mechanical Engineering*, 2020, 44(10): 22–27.

[8]   Qi P B, Liang X B, Gong Y K, Cheng Y X, Zhang Z B. Effect of milling time on mechanical alloying preparation of NbMoTaW high entropy alloy powder[J]. *Rare Metal Materials and Engineering*, 2019, 48(8): 2623–2629.

[9]   Ding P P, Mao A Q, Zhang X, et al. Preparation, characterization and properties of multicomponent AlCoCrFeNi2.1 powder by gas atomization method[J]. *Journal of Alloys and Compounds*, 2017, 721: 609–614.

[10] Yang C C, Hang Chau J L, Weng C J, Chen C S, and Chou Y H. Preparation of high-entropy AlCoCrCuFeNiSi alloy powders by gas atomization process[J]. *Materials Chemistry and Physics*, 2017, 202: 151–158.

[11] Chen C S et al. Novel cermet material of WC/multi-element alloy[J]. *International Journal of Refractory Metals and Hard Materials*, 2014, 43: 200–204.

[12] Zhou P F, Xiao D H, and Yuan T C. Comparison between ultrafine-grained WC-Co and WC-HEA-cemented carbides[J]. *Powder Metallurgy*, 2017, 60(1): 1–6.

[13] Fu Z, Raist K. TiNiFeCrCoAl high-entropy alloys as novel metallic binders for TiB2-TiC based composites[J]. *Journal of the American Ceramic Society*, 2017, 100: 7.

[14] Obra D L, Sayagués M J, Chicardi E, Torres Y, et al. A new family of cermets: Chemically complex but microstructurally simple[J]. *Journal of Alloys and Compounds*, 2020, 814: 152218.

8 High-Entropy Ceramics and Intermetallic Compounds

Yong Zhang and Yaqi Wu

8.1 INTRODUCTION

Material entropy is a thermodynamic parameter reflecting its disorder or anarchy. Substructural factors influence entropy, including magnetic moments, atomic vibrations, and atomic arrangements, while the latter is often the most effective when it comes to setting entropy. It is impossible to ignore the effects of entropy. Ceramics will play a vital role in the future of technology as new materials are developed in both bulk and thin film forms.

Recently, high-entropy alloys (HEAs) have become more popular because entropy has been underestimated in material design [1,2]. High-entropy ceramics (HECs) have been developed based on the breakthrough in metals. The stabilization of entropy in a mixture of oxides was demonstrated in 2015 [3]. In the following decades, many high-entropy disordered ceramic materials appeared, producing materials exhibiting a blend of properties that were often enhanced by adding others. In this study, we analyzed the number of high-entropy ceramics published in recent years and the proportion of each type of high-entropy ceramic (Figure 8.1).

HECs are multi-component solid solutions containing equal amounts of five or more metal cations [4]. Several recent developments in high-entropy ceramics will be discussed, such as oxides, carbides, nitrides, and intermetallic compounds, which are compounds rather than elements. Additionally, the synthesis and preparation methods, as well as the relevant applications of high-entropy ceramics, will be briefly discussed.

8.2 CLASSIFICATION OF HIGH-ENTROPY CERAMICS ACCORDING TO CHEMICAL COMPOSITION

The preparation of novel systems and initial exploration of the properties of high-entropy ceramic materials have been the focus of research on this newly emerging class of ceramic materials. As a result of their chemical composition, high-entropy

DOI: 10.1201/9781003319986-8

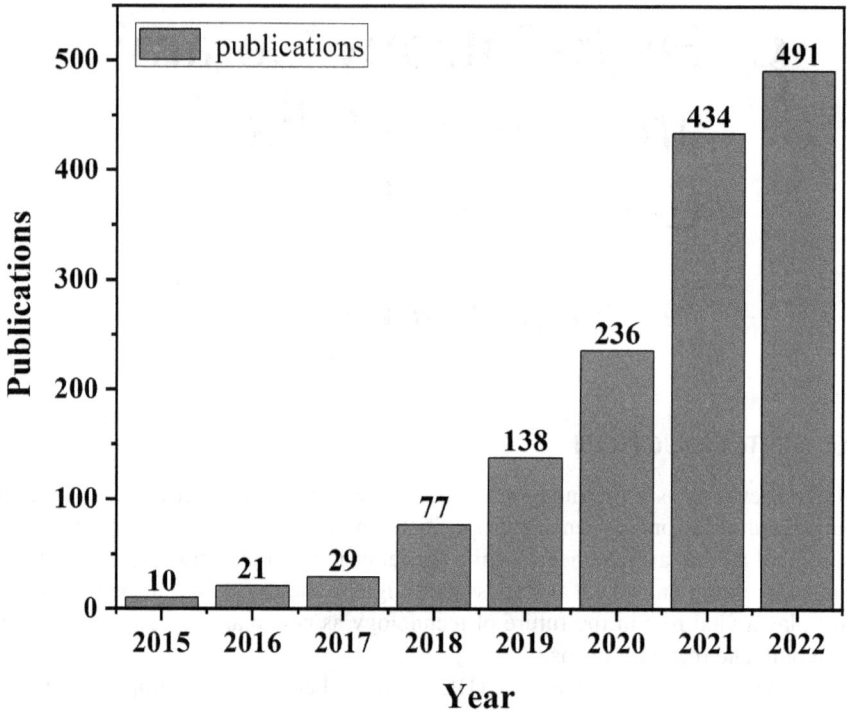

FIGURE 8.1 Statistics on the number of articles published on high-entropy ceramics in recent years, up to November 20, 2022.

ceramic materials are classified into carbides, nitrides, oxides, borides, and silicides.

8.2.1 High-Entropy Oxide of Rock Salt Structure

Using MgO, NiO, CoO, CuO, and ZnO as the initial raw materials, Rost [3], Maria, and Cutarolo et al. reported high-entropy ceramic materials in 2015 [3]. Except for CuO and ZnO, the other three oxides are rock salt structures. Single-phase (MgNiCoCuZn) O high-entropy ceramics are formed by uniformly mixing the five types of maintenance V, heating them in air, and holding them at 875°C for 12 hours. X-ray diffraction patterns and transmission electron microscopy demonstrated that the five cations are distributed randomly and uniformly in the single-phase oxides. A modified solution combustion synthesis method is currently used to manufacture HECs, along with solid-phase reactions, spray pyrolysis, flame pyrolysis, co-precipitation, and spray pyrolysis. Solid-phase reaction, spray pyrolysis, flame pyrolysis, co-precipitation, and modified solution combustion are currently employed in high-entropy ceramic synthesis. The homogeneous oxide powders cannot form a single phase under 1000°C for the solid-phase reaction method. Instead, a mixture of multiple oxides is obtained. It is also necessary to consider the cooling rate after sintering. A slow cooling rate results In the single phase easily decomposing and creating a multiphase mixture of powder at room temperature. In this case, rapid cooling is the most appropriate method for cooling.

High-entropy oxides in rock salt structures differ in their structure and properties based on the conditions of experimentation. When Cu is added to (MgCoNiCuZn)O, the anionic sublattice is significantly distorted, and the greater the Cu content, the more the coordination environment changes, which results in a more deviant structure. Cu content increases with concentration, resulting in a wider variation in the coordination environment and a more pronounced structural deviation. As Cu contents increase, the coordination environment varies more broadly and structural deviations from rock salt structure become more pronounced. As well as (MgCoNiCuZn)O, Rost et al. attempted to dope it with Li and Sr. A decrease in the band gap caused by Sr increases the disorder of the system, whereas a decrease in Li leads to a smaller lattice constant caused by Li. By solution combustion, Aqin Mao et al. [5] synthesized ultrafine powders with a particle size of 43 nm from metal nitrates and glycine. With spray pyrolysis, flame spray pyrolysis, and reverse co-precipitation, Sarkar also synthesized single-phase high-entropy oxide powders with particle sizes of only tens of nanometers.

Several functional properties of MgNiCoCuZnO have been studied due to its local disordered structure [6]. Braun's team [6] prepared high-entropy oxide ceramics (MgNiCoCuZnA)O with low thermal conductivity similar to amorphous materials and a high modulus. According to Brardan [7] et al., certain (MgNiCoCuZn)-based high-entropy ceramics have oxygen vacancies that result in materials with exceptionally high ion transport rates. As catalysts for CO oxidation, $(Mg_{0.2}Co_{0.2}Ni_{0.2}Zn_{0.2}Cu_{0.2})O$ high-entropy oxygenates were studied by Chen et al. [8] They found that $(Mg_{0.2}Co_{0.2}Ni_{0.2}Zn_{0.2}Cu_{0.2})O$ had a high catalytic activity, and the catalytic process was stable at high temperatures due to its high-temperature stability.

8.2.2 Fluorite Structure High-Entropy Oxide

The term fluorite structure high-entropy oxide refers to materials with a fluorite structure composed of three or more cations. It was reported in 2017 that fluorite-type high-entropy oxides could be prepared, such as (CeLaNdSmY)O, (CeLaPrSmY)O, (GdLaNdPrSmY)O, and (CeGdLaNdPrSmY)O [9,10] using spray pyrolysis. Adding rare earth elements to high-entropy oxides increases the oxygen vacancy concentration. Oxides' band gaps can be reduced by Pr. The Ce ion plays a crucial role in phase stability, and different types of elements and synthesis processes will affect the stability of materials. It has been reported that Gild et al. [11] prepared fluorite-structured high-entropy ceramics with a lower electrical and thermal conductivity than yttrium oxide stabilized zirconia (8 YSZ), which may be due to higher phonon scattering. As previously reported by Chen et al., [12] at room temperature, $(Zr_{0.2}Ce_{0.2}Hf_{0.2}Ti_{0.2}Sn_{0.2})O_2$ fluorite ceramic has a thermal conductivity of 1.28 W/(m·K).

As well as high-entropy oxide ceramics (HEOCs), non-oxide high-entropy ceramics are also receiving a lot of attention. A high melting point and hardness make non-metals like B and C useful in refractory materials. Boride ceramics have a high degree of covalent bonding because they are generally hexagonal in shape,

with B atoms forming the lamellar structure and metal atoms occupying the interlayer positions. In addition to borides' excellent properties, the solid solution of
various elements allows the design of high-entropy materials with specific properties. It has been reported that Gild has prepared $(Ti_{0.2}Zr_{0.2}Nb_{0.2}Hf_{0.2}Ta_{0.2})B_2$ and six
other single-phase HEBCs in China for the first time. [13] As a result of the high
impurity content in the powder, the sintered ceramic body is not dense enough.
Adding a part of C in the later stage helps to reduce the problem of having too
low density, but oxygenated impurities and incomplete C remain in the material.
A high-entropy boride prepared from a single component has higher hardness and
stronger resistance to oxidation. Using metal oxide and amorphous boron powder
as raw materials, Zhang et al. [14] synthesized $(Ti_{0.2}Cr_{0.2}Zr_{0.2}Hf_{0.2}Ta_{0.2})B_2$, $(Ti_{0.2}Zr_{0.2}Nb_{0.2}Mo_{0.2}Hf_{0.2})B_2$, and other submicron powders using boron thermal reduction
[14]. The sintered powder has a maximum relative density of 99.2%, a maximum
hardness of 26 GPa, and still contains 2.4%~7.2% of impurities. $(Ti_{0.2}Nb_{0.2}Hf_{0.2}Mo_{0.2}Ta_{0.2})B_2$ powder was first prepared by Tallarita et al. [15]. Even though sintered powder does not increase the density of the material, its relative density is
92.4%. In spite of this, this method reduces the ball milling time and eliminates
oxide impurities because there is no oxide raw material. Gu et al. [16] obtained a
submicron five-element high-entropy boride $(Ti_{0.2}Hf_{0.2}Zr_{0.2}Nb_{0.2}Ta_{0.2})$. Based on an
X-ray diffraction analysis of B_2 powder, a single-phase material is obtained, but
the element distribution is not uniform when boron thermal reduction is performed
at 1800 °C. When the reaction temperature is raised to 2000 °C, a single-phase
material can be obtained with a uniform element distribution. High-entropy boride
ceramics also improve their mechanical properties with a temperature increase
from 2000°C to 2050°C.

8.3 CARBIDES

HECCs are currently being studied in a similar manner to HECCs, and they primarily consist of transition metal carbides of the IVB and VB groups. HECCs have
a rock salt structure, and since carbides and borides have covalent bonding properties, these carbides and borides have high melting points as well. They can be used
under extreme conditions and have expanded the range of carbide ceramics. In the
early years of the study of transition refractory metal carbide ceramics, the focus
of the research was on the five-element (TiZrHfNbTa)C system. However, it gradually spread to the four-element (TiZrNbTaMo)C, the five-element (TiZrNbTaW)
C, and the transition refractory metal carbide ceramics, followed by the five-
element system (TiZrHfVNb)C. Castle [17] synthesized high-entropy carbide
ceramics (TiZrHfTa)C and (ZrNbHfTa)C for the first time in 2018. In contrast,
electron backscattering (EBSD) and Cs-STEM techniques found that (TiZrHfTa)
C was uniformly distributed, whereas (ZrNbHfTa)C was not. Therefore, Nb is
more suitable than Ti for preparing high-entropy carbonized ceramics of ZrHfTa.
Harrington [18] calculated 12 high-entropy carbide ceramics by using the entropy
formation principle operator based on the first principles proposed by Sarker [19].
The mechanical properties of nine of the high-entropy carbides were superior to
those of the binary carbides, as observed experimentally. A high-entropy ceramic

$(Hf_{0.2}Zr_{0.2}Ta_{0.2}Nb_{0.2}Ti_{0.2})C$ was prepared by Ye et al. [20] in 2019. It has a wide range of applications for harsh environments based on testing its mechanical properties. Researchers [21] have also produced single-phase (TiZrNbTaW)C ceramics using three different sintering methods: carbonized mutual solid solution, elemental reaction sintering, and redox sintering. Oxide reduction sintering produced carbide ceramics with a relatively high density, but partial Zr distributions were observed. Furthermore, high-entropy carbides are gradually being investigated in terms of their properties, along with their preparation [22]. High-entropy carbide ceramics have been extensively researched by related researchers. The researchers measured the stiffness, hardness, thermal conductivity, thermal stability, and oxidation resistance of high-entropy carbide ceramics. However, ceramic materials perform better when their densities are higher, and increasing their densities can help. A ternary high-entropy carbide (HfZrTa)C was produced by Zhang et al. [23] using the arc melting method. Ball milling-discharge plasma sintering (SPS) was used by Zhou [24] to produce high-entropy alloy powders with an isoatomic ratio $(Ti_{0.2}Zr_{0.2}Hf_{0.2}Nb_{0.2}Ta_{0.2})C$. Using small steel balls, Chicardi E et al. [25] synthesized $Ti_{0.2}Zr_{0.2}Hf_{0.2}V_{0.2}Nb_{0.2}C$ HECCs powder. Single-phase HECCs can be formed during ball milling due to the reduction of impact energy [26]. According to Chen [27], high-entropy porous carbide ceramics with a porosity of 80.99% were developed based on the low thermal conductivity high-entropy ceramics studied by Yan [28]. Moreover, they obtained high-entropy carbide ceramics with a thermal conductivity of only 5% of dense ceramics.

8.4 NITRITES

Nitrides are also one of the most influential ceramic materials. In the early days of high-entropy materials, nitrides were one of the first materials to be reported to have high entropy. In comparison to binary, ternary, and quarterly nitride materials, HENs possess significantly superior mechanical properties, thermal stability, corrosion resistance, and wear resistance, in comparison to binary, ternary, and quarterly nitride materials. Using reactive sputtering, Chen et al. [29] fabricated FeCoNiCrCuAlMn and $FeCoNiCrCuAl_{0.5}$ nitrides and investigated their hardness as coatings [29]. Many studies have focused on the mechanical properties of HENs and their resistance to corrosion and oxidation since 2004, when HENs were considered promising materials for protective coatings.

A reactive magnetron sputtering coating of two HENs was synthesized by Hsieh et al. [30] and investigated the impact of substrate bias on the chemical composition, microstructure, and mechanical properties of the coatings. By increasing the substrate bias, the hardness of both HENs increased to 36.1–36.7 GPa [31]. As reported by Kirnbauer et al. [32] the (Hf, Ta, Ti, V, Zr)N film synthesized by reactive magnetron sputtering with flow rates of $N_2/(Ar + N_2)$ of 0.3 and 0.4 has a FCC structure with a hardness value of 32.5 ± 0.8 GPa. As a result of annealing at 1573 K, the hardness was only slightly reduced and reached 28.1 ± 1.4 GPa, while the crystal structure remained stable. Cui et al. [33] studied the effect of the nitrogen content on the microstructures and mechanical properties of (AlCrTiZrHf)N. The researchers demonstrated that HEN changes to amorphous structures under increased nitrogen flow, resulting in a 33.1 GPa hardness and an elastic modulus of 347.3 GPa. Increased solid-solution strength was attributed to the formation of a saturated nitride phase. The interlayer of $(Al_{0.34}Cr_{0.22}Nb_{0.11}Si_{0.11}Ti_{0.22})_{50}N_{50}$ showed superior adhesion to Ti, superior oxidation resistance, and superior thermal stability [34]. Compared to inserts coated with TiN and TiAlN, inserts coated with this HEN coating and Ti interlayer showed a fine milling behavior, indicating that HEN-Ti is a potential protective coating. A high-entropy effect, severe lattice distortion, and low grain boundary energy all contributed to the remarkably high mechanical stability of this HEN [35].

A protective coating that is resistant to oxidation was developed for several HENs. In an investigation of the oxidation behavior of $(Al_{0.34}Cr_{0.22}Nb_{0.11}Si_{0.11}Ti_{0.22})_{50}N_{50}$, Shen et al. [36] found that the resistance to oxidation is high up to 1573 K. During oxidation at 1573 K for 50 hours, only 0.015 mgcm^{-2} of weight was gained. Various oxide layers were found on the surface, each with a different composition. This HEN high oxidation resistance was attributed to its Al_2O_3 surface layer and Si-rich amorphous networks. Pogrebnjak et al. [37] used cathodic-arc-vapor deposition to study the oxidation behavior and tribological features of films made from (TiHfZrVNb)N. Rutile TiO_2 and other $MTiO_4$-type oxides were formed by oxidation of HEN at 873 K. HEN coatings $(Al_{29.1}Cr_{30.8}Nb_{11.2}Si_{7.7}Ti_{21.2})N_{50}$ and $Al_{23.1}Cr_{30.8}Nb_{7.7}Si_{7.7}Ti_{30.7})N_{50}$ were oxidized at 1173 K for 2 h in air by Hsieh et al. [30] There was a noticeable difference in the thickness of the oxide layer for $Al_{23.1}Cr_{30.8}Nb_{7.7}Si_{7.7}Ti_{30.7}N_{50}$ and for $Al_{29.1}Cr_{30.8}Nb_{11.2}Si_{7.7}Ti_{21.2}N_{50}$, ranging from 80 nm to 7 nm. Despite its higher Al/Ti ratio (Al_2O_3 layer is more protective than TiO_2), $(Al_{29.1}Cr_{30.8}Nb_{11.2}Si_{7.7}Ti_{21.2})N_{50}$ is still more oxidation resistant than $(Al_{29.1}Cr_{30.8}Nb_{11.2}Si_{7.7}Ti_{21.2})N_{50}$ [38–42]. TiN coatings have excellent anti-corrosion properties due to the formation of a nitrogen-rich layer on the surface [41,43]. A few studies have evaluated the corrosion resistance of HENs, suggesting reasonable levels of resistance. Based on the results of Hsueh et al. [38], the nitrogen fraction and bias of $(AlCrSiTiZr)_{100-x}N_x$ films were studied. Anodic polarization analysis of corrosion behavior under ambient conditions was used to study corrosion behavior. Changing the film composition, film structure, film density, and coat-substrate bonding can significantly improve corrosion resistance by tweaking the nitrogen content and increasing the substrate bias. In addition to (Al, Cr, Nb, Y, Zr)N, Fieandt et al. [44] showed that the material was more corrosion resistant than hyperduplex stainless steel and ternary thin films (Nb, Zr)N, which were commercially available.

Researchers have also studied high-entropy non-oxide ceramics, such as high-entropy nitrides and silicides, in addition to high-entropy oxides, borides, and carbides. A substantial number of research has been conducted on these non-oxide ceramics, and it is inferred that, with gradual improvements, they will have many applications in the future. In 1945, Gild [45] and Qin et al. [46] proposed the concept of silicified ceramics. Discharge plasma sintering was used to prepare the $(TiNbTaMoW)Si_2$ ceramics. With a Vickers hardness of 12 GPa and a thermal conductivity of less than 1 W/(m·K), the prepared ceramic crystals have a hexagonal dense row structure and favorable mechanical properties. In order to produce ceramics with 99% density $(TiNbTaMoW)Si_2$, metals and silicon elements were used as raw materials. There was, however, no clear reason why the $(TiNbTaMoW)Si_2$ ceramics prepared by both had different degrees of Zr elemental segregation, as Chen [47]pointed out. Hsieh [48] $(Al_{23.1}Cr_{30.8}Nb_{7.7}Si_{7.7}Ti_{30.7})N_{50}$ and $(Al_{29.1}Cr_{30.8}Nb_{11.2}Si_{7.7}Ti_{21.2})N_{50}$ are two high-entropy nitride coatings, both with NaCl-type structures and excellent oxidation resistance. Based on first-principle calculations, Wen [49] synthesized high-entropy aluminum silicates $(Mo_{0.25}Nb_{0.25}Ta_{0.25}V_{0.25})(Al_{0.5}Si_{0.52})$(HEAS-1). They first performed calculations to determine whether or not HEAS-1 could be generated, mainly in terms of thermodynamic and lattice size differences, and we were able to generate HEAS-1 using solid-phase reactions.

8.5   SYNTHESIS AND PREPARATION METHODS

Since HECs have a high melting point, they are much more difficult to fabricate than conventional metals. Furthermore, they can be prepared in a wide variety of ways. This means that there are many ways to classify the ways in which HECs are prepared. The preparation of HECs can be divided into three categories based on the three-dimensional scale of the resulting samples: powder preparation, thin films or thick coating preparation, and bulk alloy preparation. In light of other categories, the synthesis methods for HECs can be classified into three main groups: solid-state methods, liquid-state methods, and gas-state methods.

In this section, the main synthesis methods to produce HECs will be reviewed from the perspective of three-dimensionality.

8.5.1   POWDER

In addition to anodes of Li-ion batteries, catalysts, electromagnetic wave absorbers, and others, high-quality HEC powders are filling materials used in sintering bulk materials. Various methods are available for preparing high-entropy ceramics, such as solid-state reactions [50–58], wet chemical routes [59–64], solidification [65–69], field-assisted synthesis [70,71], and epitaxial growth. Powder preparation from different HEC systems is traditionally achieved through solid-phase reactions (solid-state reaction, SSR). Spinel-structured $(CoCrFeMnNi)_3O_4$ was prepared by vibration milling Co_3O_4, Cr_2O_3, Fe_2O_3, MnO, and NiO powders and then calcining them for 20 hours at 1050 °C [72]. The powder mixtures of Na_2CO_3, Bi_2O_3, $BaCO_3$, $SrCO_3$, $CaCO_3$, and TiO_2 were calcined at 1050°C by Pu et al. [73]. A high-entropy rare earth silicide carbide was synthesized at 1600°C by Chen et al. [54].

The sol-gel method can also be used to prepare high-entropy ceramic powders. They used the sol-gel method [74] to synthesize $(Yb_{0.2}Y_{0.2}Lu_{0.2}Sc_{0.2}Gd_{0.2})_2Si_2O_7$ powders. Transition metal carbides and borides require very high sintering temperatures. To increase the density of the prepared product, it is necessary to reduce the surface energy. To achieve such a goal, various methods have been used to synthesize ultrafine high-entropy carbide and boride powders. Through reactive high-energy ball milling (R-HEBM) of elemental transition metal powders and graphite particles, Moskovskikh et al. [75] synthesized high-entropy powders $Hf_{0.2}Ta_{0.2}Ti_{0.2}Nb_{0.2}Zr_{0.2}C$

and $Hf_{0.2}Ta_{0.2}Ti_{0.2}Nb_{0.2}Mo_{0.2}C$. According to Guan et al. [76], mechanical alloying at room temperature produced high-entropy metal boron carbonitride powders $(Ta_{0.2}Nb_{0.2}Zr_{0.2}Hf_{0.2}Ti_{0.2}BCN)$ and $(Ta_{0.2}Nb_{0.2}Zr_{0.2}Ti_{0.2}W_{0.2})BCN$.

A nanometer-sized powder of $(Ta_{0.25}Nb_{0.25}Ti_{0.25}V_{0.25})C$ was prepared by the molten salt method as well. Ning et al. synthesized high-entropy $(Ta_{0.25}Nb_{0.25}Ti_{0.25}V_{0.25})$ C from Ta, Nb, Ti, and V powders at 1300 °C using KCl as the molten salt medium [77]. For the synthesis of fine-grained, high-quality powders, carbothermal reduction, borothermal reduction, and boro/carbothermal reactions are also advantageous. As an example, Feng et al. [78] synthesized $(Hf_{0.2}Zr_{0.2}Ti_{0.2}Ta_{0.2}Nb_{0.2})C$ powders with an average particle size of about 550 nm at 2000 °C by combining transition-metal oxides and carbon powders. By utilizing a facile borothermal reduction method, Liu et al. [79] produced $(Hf_{0.2}Zr_{0.2}Ta_{0.2}Nb_{0.2}Ti_{0.2})B_2$ with 310 nm particles at 1700 °C. As reduction agents, Zhang et al. [80] used B_4C and graphite powders to prepare $(Hf_{0.2}Zr_{0.2}Ta_{0.2}Nb_{0.2}Ti_{0.2})B_2$, $(Hf_{0.2}Zr_{0.2}Mo_{0.2}Nb_{0.2}Ti_{0.2})B_2$, and $(Hf_{0.2}Mo_{0.2}Ta_{0.2}Nb_{0.2}Ti_{0.2})B_2$ powders. A near-fully dense bulk carbide and boride HEC with improved mechanical properties were prepared using ultrafine high-entropy carbides and borides as initial materials [75,80–82].

8.5.2 BULK

For the preparation of bulk ceramic materials, mechanical alloying, electrical discharge plasma sintering, and hot pressing sintering are common methods. Nevertheless, the current method of preparing high-entropy ceramics has some disadvantages: (1) due to the introduction of mechanical alloying methods, ball-milled powders often contain impurities brought in by the grinding balls and jars, and sintered ceramics are prone to uneven element distribution; (2) hysteresis diffusion makes the densification of high-entropy ceramics difficult, resulting in small pores that interfere with subsequent performance testing; (3) ceramics contain a small amount of oxygen when oxides are used for carbonization. The synthesis of ceramic powders will be improved in the future in order to improve the relative density of sintered ceramics and reduce the content of impurities.

A recent study combined flash sintering with spark plasma sintering to improve synthesis densification. Sintering with spark plasma is becoming increasingly popular for the production of bulk high-entropy materials. This method, referred to as reactive flash spark plasma sintering, was used to synthesize $(Hf_{0.2}Zr_{0.2}Ta_{0.2}Nb_{0.2}Ti_{0.2})$ B_2 [83] and $(Hf_{0.2}Zr_{0.2}Ta_{0.2}Nb_{0.2}Ti_{0.2})C$ [83]. In the study conducted by Zhang et al. [84], high-entropy powders synthesized by borothermal reduction were used as initial materials in the preparation of $(Hf_{0.2}Zr_{0.2}Ta_{0.2}Cr_{0.2}Ti_{0.2})B_2$, $(Hf_{0.2}Mo_{0.2}Zr_{0.2}Nb_{0.2}Ti_{0.2})$ B_2, and $(Hf_{0.2}Mo_{0.2}Ta_{0.2}Nb_{0.2}Ti_{0.2})_2$. $(Ti_{0.2}Hf_{0.2}Zr_{0.2}Nb_{0.2}Ta_{0.2})B_2$ [85] was synthesized by SPS at 2000–2050°C and 50 MPa of boro/carbothermal reduction. A two-step SPS [86] process was used by Feng et al. [87] to prepare dense $(Hf_{0.2}Zr_{0.2}Ti_{0.2}Ta_{0.2}Nb_{0.2})B_2$. The boro/carbothermal solid was heated at 1650°C in a mild vacuum at 15 MPa load in the first step, and densification was carried out at 2000–2200°C under 50 MPa in the second step. A high-entropy carbide, $(Hf_{0.25}Ta_{0.25}Zr_{0.25}Ti_{0.25})C$, and $(Hf_{0.25}Ta_{0.25}Zr_{0.25}Nb_{0.25})C$ were prepared by SPS at 2300°C and 16–40 MPa under 30 MPa pressure by Castle et al. [88]. The ceramics made by Sarker et al. [89] had a relative density of 99% when manufactured by SPS at 2200°C.

Ceramics with porous structures play a crucial role in ultrahigh-temperature insulation. For fabricating highly porous bulk materials, low thermal conductivity and high-entropy carbides and borides were selected as the backbone materials, and a novel in situ reaction/partial sintering method was developed [90–92]. This novel method has been used to prepare high porosity $(Zr_{0.2}Hf_{0.2}Ti_{0.2}Nb_{0.2}Ta_{0.2})C$ [93], $(Zr_{0.2}Hf_{0.2}Ti_{0.2}Nb_{0.2}Ta_{0.2})B_2$ [94], and $(Y_{0.2}Yb_{0.2}Sm_{0.2}Nd_{0.2}Eu_{0.2})B_6$ [86]. In a similar manner, porous $(Zr_{0.2}Hf_{0.2}Ti_{0.2}Nb_{0.2}Ta_{0.2})B_2$ with a porosity of 75.7%, grain size of 400–800 nm, porosity of 0.3–1.2um, compressive strength of 3.93 MPa, and thermal conductivity of 0.51 $W \cdot m^{-1} \cdot K^{-1}$ was prepared.

8.5.3 THIN FILMS AND THICK COATINGS

A sputtering process involves bombarding a solid (target) with accelerated charged ions, causing the surface atoms to scatter backward, a physical vapor deposition (PVD) method. In thin film production, the sputtering process is used to deposit thin films. As a conventional method of producing thin film materials, plasma deposition involves a substrate, such as a silicon wafer, and a target, an alloy of the desired elements. Both are placed in a vacuum chamber. Gases such as Ar are used to bombard the target and remove the elements by sputtering. In order to produce HENs, HEOs, HEBs, HEONs, and HECs, sputtering techniques include ion beam sputtering, reactive sputtering, diode sputtering, radiofrequency sputtering, and magnetron sputtering.

The process of pulsed laser deposition is one of the easiest ways to create thin films. It involves ablating the material from a target surface with a high-power laser. This method produces ferroelectric, ferromagnetic, and dielectric oxides. Ablation of the material is accomplished by focusing a pulsed laser on the target surface. An intense pressure plasma is generated by high temperatures and ablation, which then expands locally and directionally to deposit a film on the substrate. A vacuum or oxygen is also an option for Meisenheimer et al. [95] when creating oxide films. Kotsonis et al. [96] used the method and produced $Mg_xNi_xCo_xCu_xZn_xSc_xO$ and $Mg_xNi_xCu_xZn_xSc_xO$, respectively.

8.6 PROPERTIES AND POTENTIAL APPLICATIONS

As structural and functional materials, ceramics primarily contain ionic or covalent bonds. Due to their high hardness, melting temperature, and oxidation and corrosion resistance, they are used as coatings or high-temperature materials. Ceramics can be utilized in the electrical and electronic industries because they have different electronic properties than metallic materials, which are usually conductors. Due to the cocktail effect, electron distribution, lattice distortion, and slow diffusion, high-entropy ceramics are usually more stable than conventional ceramics with interesting properties [97–114]. In this section, the properties of high-entropy ceramics, including HEOs, HENs, HECs, HEBs, and HEHs, are reviewed, and their potential applications are discussed.

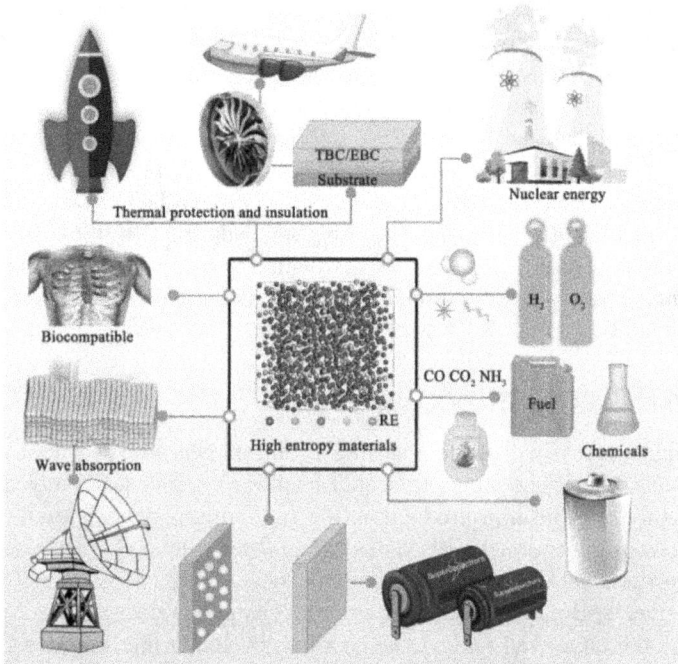

FIGURE 8.2 Prospective applications of HECs covering ultrahigh temperature structural to energy and catalytic functional applications.

Source: Open Access; refer to Xiang, H., Xing, Y., Dai, Fz. et al. High-entropy ceramics: Present status, challenges, and a look forward. J Adv Ceram 10, 385–441 (2021). https://doi.org/10.1007/s40145-021-0477-y.

8.6.1 High-Entropy Ceramic Coatings

High-entropy ceramic materials can experience a reduction in thermal conductivity due to the chaotic distribution of multiple (five or more) ions (usually positive ions) in equivalent lattice positions [117]. The work of Professor Luo involved the preparation of a number of high-entropy or medium-entropy fluorite oxide ceramics with oxygen vacancies, which have low thermal conductivity. The thermal conductivity is 1.10–1.81 W/(m·K) [118], which is significantly lower than the thermal conductivity of the typical thermal barrier coating material 8YSZ. In addition to being potential thermal barrier coating materials, high-entropy oxides can also improve thermoelectric materials due to the reduction in thermal conductivity. Entropy engineering was proposed by Chen. It has been demonstrated that configurational entropy plays a significant role in the formation of multicomponent thermoelectric materials, such as $(Cu/Ag)(In/Ga)Te_2$ and $(Mn/Ge/Sn/Pb)$ Te-based single-phase materials. An analysis of performance tests and theoretical predictions shows that high configurational entropy leads to the formation of single phases. The decreasing of lattice thermal conductivity and increasing Seebeck coefficients lead to an increase in room-temperature thermoelectric optimum values [119]. In similar studies, it has also been demonstrated that increasing the entropy of the system with $AgSbSe_2$ or $AgBiSe_2$ to GeSe promotes the conversion of the rhombic phase to the cubic phase in GeSe and, therefore, improves thermoelectric performance [120,121].

Solid-solution strengthened carbides, nitrides, carbonitrides, and oxides are more durable due to the sluggish diffusion of oxygen and other species, and they are also more resistant to oxidation and corrosion. For machine parts, such as cutting tools and drill bits, they are valuable as protective coatings resistant to wear, oxidation, and/or corrosion. These parts operate at high temperatures in corrosive environments. Several properties of carbide, nitride, and oxide coatings were measured, including hardness, elastic modulus, corrosion current and potential in corrosive solutions, and wear rate. Due to the lower coefficient of friction of carbides, nitrides can be more durable than carbides. Free carbon layers acting as lubricants on the surface can explain this [122]. Among oxide coatings, $Al_2(CoCrCuFeNi)O$ and $(AlCrTaTiZr)O$ exhibit the highest hardness values; binary oxides rarely surpass 20 GPa (Cr_2O_3 does with 25–30 GPa) [123,124].

As a result of higher aluminum and oxygen concentrations [125] and annealing, hardness generally increases with substrate bias and nitrogen content [123]. It is explained by the incorporation of strong oxygen-metal bonds, the elimination of film cracks, and the formation of nanocrystalline phases [124].

8.6.2 Thermoelectrics

Thermoelectricity is always being studied in order to improve its conversion efficiency. Energy generation and refrigeration are made possible by thermoelectric materials. They are evaluated by the figure of merit: $zT = \sigma S2/\kappa$, where σ is the electrical conductivity, S is the Seebeck coefficient, and κ is the thermal conductivity—the reduced lattice thermal conductivity in disordered materials leads to increased zT. Perovskite-structure $(BaCaLaPbSr)TiO_3$ had a thermal conductivity five times smaller than $SrTiO_3$ [126], with a maximum zT over 0.2. The high-entropy chalcogenide $Pb_{0.89}Sb_{0.012}Sn_{0.1}Se_{0.5}Te_{0.25}S_{0.25}$ had an ultralow thermal conductivity of 0.3 W m^{-1}K^{-1} [127] with zT of 1.8 compared to 0.8 for $Pb_{0.99}Sb_{0.012}Se$. The addition of Sn resulted in increased configurational entropy, stabilizing the Pb(SeTeS) system. $Pb_{0.975-x}Cd_xNa_{0.025}Se_{0.5}S_{0.25}Te_{0.025}$, where $x \leq 0.05$, had a power output of 2.7 W with a conversion efficiency of 12% (among the highest reported values) [125].

8.6.3 Catalysts

It is well known that high-entropy ceramics can be used as catalysts, electrocatalysts, and photocatalysts, as well as carriers for catalysts. [128]. For catalysis, high-entropy metal alloys have been studied for a while [129] and have been shown to possess useful qualities, including corrosion resistance, which limits traditional transition metal alloys as electrocatalysts in acidic or alkaline environments [130]; resistance to poisoning; a large number of unique binding sites providing a nearly continuous distribution of adsorption energies; synergetic and/or unexpected activity enhancements; and increased miscibility of elements allowing the optimization of binding strength for higher activity and increased miscibility of elements, allowing the optimization of binding strength for higher activity.

Several reactions have been demonstrated to be effective with high-entropy alloys, including oxidation (methanol [131–133], ammonia [134], carbon monoxide [133]), evolution (oxygen [135], hydrogen [130]), decomposition (ammonia [136]), reduction (oxygen [133,137–139]), and degradation (azo dye). It is possible to catalyze reactions in high-entropy ceramics by combining bond chemistries. The following three reports are relevant. A group of researchers investigated (MgCoNiCuZn)O modified with precious metals for the oxidation of CO [140] and hydrogenation of CO_2 [141]. Up to 5 wt.% of disordered oxide was found to enhance Pt/Ru dispersion and activity, and to resist precious metal sintering during high-heat treatments (900°C for CO oxidation and 700°C for CO_2 hydrogenation). CO was completely converted at 155°C on the Pt-loaded catalyst, which retained high reusability at the reaction temperature. Catalysts loaded with 5 wt.% Pt/Ru showed over 45% CO yields at 500 °C and over 40% CO_2 conversions. The CO selectivity of Pt/Ru-loaded CO_2 hydrogenation catalysts was over 95% [141]. The second set of reports investigated the oxidation of CO in mesoporous $(MgCoNiCuFe)O_x–Al_2O_3$. CO conversion is complete at 260°C. Compared to $CuO–Al_2O_3$, the ceramic exhibited negligible degradation after 48 h [142]. Additionally, nitrogen, carbon, oxygen, and boron systems were identified as promising oxygen reduction systems in the third set of reports [137,143–149].

REFERENCES

[1] Yeh JW, Chen SK, Lin SJ, et al. Nanostructured high-entropy alloys with multiple principle elements: Novel alloy design concepts and outcomes. *Advanced Engineering Materials*, 2004, 6, 299–303.

[2] Cantor B, Chang ITH, Knight P, et al. Microstructural development in equiatomic multicomponent alloys. *Materials Science and Engineering A*, 2004, 375–377, 213–218.

[3] Rost CM, Sachet E, Borman T, et al. Entropy-stabilized oxides. *Nature Communications*, 2015, 6, 8485.

[4] Toher C, Oses C, Hicks D, et al. Unavoidable disorder and entropy in multi-component systems[J]. *NPJ Computational Materials*, 2019, 5, 69.

[5] Mao A, Xiang H-Z, Zhang Z-G, et al. Solution combustion synthesis and magnetic property of rock-salt (Co0.2Cu0.2Mg0.2Ni0.2Zn0.2)O high-entropy oxide nanocrystalline powder. *Journal of Magnetism and Magnetic Materials*, 2019, 484, 245–252.

[6] Braun JL, Rost CM, Lim M, et al. Charge-Induced disorder controls the thermal conductivity of entropy-stabilized oxides[J]. *Advanced Materials*, 2018, 30, 1805004.

[7] Bérardan D, Franger S, Meena AK, et al. Room temperature lithium superionic conductivity in high entropy oxides. *Journal of Materials Chemistry A*, 2016, 4(24), 9536–9541.

[8] Chen H, Fu J, Zhang P, et al. Entropy-stabilized metal oxide solid solutions as CO oxidation catalysts with high-temperature stability. *Journal of Materials Chemistry A*, 2018, 6(24), 11129–11133.

[9] Djenadic R, Sarkar A, Clemens O, et al. Multicomponent equiatomic rare earth oxides. *Materials Research Letters*, 2017, 5(2), 102–109. https://doi.org/10.1080/21663831.2016.1220433.

[10] Sarkar A, Loho C, Velasco L, et al. Multicomponent equiatomic rare earth oxides with a narrow band gap and associated praseodymium multivalency[J]. *Dalton Transactions*, 2017, 46, 12167–1217.

[11] Gild J, Samiee M, Braun JL, et al. High-entropy fluorite oxides. *Journal of the European Ceramic Society*, 2018, 10(38), 3578–3584.

[12] Chen K, Pei X, Tang L, et al. A five-component entropy-stabilized fluorite oxide. *Journal of the European Ceramic Society*, 2018, 38(11), 4161–4164.

[13] Gild J, Zhang Y, Harrington T, et al. High-entropy metal diborides: A new class of high-entropy materials and a new type of ultrahigh temperature ceramics. *Scientific Reports*, 2016, 6, 37946. https://doi.org/10.1038/srep37946.

[14] Zhang Y, Guo W-M, Jiang Z-B, et al. Dense high-entropy boride ceramics with ultra-high hardness. *Scripta Materialia*, 2019, 164, 135–139. https://doi.org/10.1016/j.scriptamat.2019.01.021.

[15] Giovanna T, Roberta L, Sebastiano G, et al. Novel processing route for the fabrication of bulk high-entropy metal diborides[J]. *Scripta Materialia*, 2019, 158, 100–104.

[16] Gu J, Zou J, Sun S-K, et al. Dense and pure high-entropy metal diboride ceramics sintered from self-synthesized powders via boro/carbothermal reduction approach. *Science China Materials*, 2019, 62(12), 898–1909. https://doi.org/10.1007/s40843-019-9469-4.

[17] Elinor C, Csanádi T, Salvatore G, et al. Processing and properties of high-entropy ultra-high temperature carbides[J]. *Scientific Reports*, 2018, 8(1), 8609.

[18] Harrington TJ, Gild J, Sarker P, et al. Phase stability and mechanical properties of novel high entropy transition metal carbides[J]. *Acta Materialia*, 2018, 166.

[19] Sarker P, Harrington T, Toher C, et al. High-entropy high-hardness metal carbides discovered by entropy descriptors. *Nature Communications*, 2018, 9, 4980.

[20] Ye B, Wen T, Huang K, et al. First-principles study, fabrication, and characterization of (Hf0.2Zr0.2Ta0.2Nb0.2Ti0.2)C high-entropy ceramic[J]. *Journal of the American Ceramic Society*, 2019, 7, 102.

[21] Wei XF, Liu JX, Li F, et al. High entropy carbide ceramics from different starting materials[J]. *Journal of the European Ceramic Society*, 2019, 39(10), 2989–2994.

[22] Li G, Liu N, Zhang X. Research status of high entropy carbide powders[J]. *Cemented Carbide*, 2020, 37(2), 10.

[23] Zhang Z, Fu S, Aversano F, et al. Arc melting: A novel method to prepare homogeneous solid solutions of transition metal carbides (Zr, Ta, Hf)[J]. *Ceramics International*, 2019, 45(7), 9316–9319.

[24] Zhou J, Zhang J, Zhang F, et al. High-entropy carbide: A novel class of multicomponent ceramics[J]. *Ceramics International*, 2018, 17, 44.

[25] Chicardi E, Garcia-Garrido C, Gotor FJ. Low temperature synthesis of an equiatomic (TiZrHfVNb)C-5 high entropy carbide by a mechanically-induced carbon diffusion route[J]. *Ceramics International*, 2019, 45(17 Pt.A), 21858–21863.

[26] Feng L, Fahrenholtz WG, Hilmas GE, et al. Synthesis of single-phase high-entropy carbide powders[J]. *Scripta Materialia*, 2019, 162, 90–93.

[27] Chen H, Xiang H, Dai FZ, et al. High entropy (Yb0.25Y0.25Lu0.25Er0.25)2SiO5 with strong anisotropy in thermal expansion[J]. *Journal of Material Science and Technology*, 2020, 36(1), 6.

[28] Yan X, Constantin L, Lu Y, et al. (Hf0.2Zr0.2Ta0.2Nb0.2Ti0.2)C high-entropy ceramics with low thermal conductivity. *Journal of the American Ceramic Society*, 2018, 101, 4486–4491. https://doi.org/10.1111/jace.15779.

[29] Chen T, Shun T, Yeh J, et al. Nanostructured nitride films of multi-element high-entropy alloys by reactive DC sputtering. *Surface and Coatings Technology*, 2004, 188, 193–200.

[30] Hsieh MH, Tsai MH, Shen WJ, et al. Structure and properties of two Al–Cr–Nb–Si–Ti high-entropy nitride coatings. *Surface and Coatings Technology*, 2013, 221, 118–123.

[31]   Huang PK, Yeh JW. Effects of nitrogen content on structure and mechanical properties of multi-element (AlCrNbSiTiV)N coating. *Surface and Coatings Technology*, 2009, 203, 1891–1896.

[32]   Kirnbauer A, Kretschmer A, Koller CM, et al. Mechanical properties and thermal stability of reactively sputtered multi-principal-metal Hf-Ta-Ti-V-Zr nitrides. *Surface and Coatings Technology*, 2020, 389, 125674.

[33]   Cui P, Li W, Liu P, et al. Effects of nitrogen content on microstructures and mechanical properties of (AlCrTiZrHf)N high-entropy alloy nitride film[J]. *Journal of Alloys and Compounds*, 2020, 834, 155063.

[34]   Shen WJ, Tsai MH, Yeh JW. Machining performance of sputter-deposited (Al0.34C r0.22Nb0.11Si0.11Ti0.22)50N50 high-entropy nitride coatings[J]. *Coatings*, 2015, 5, 312–325.

[35]   Huang PK, Yeh JW. Inhibition of grain coarsening up to 1000°C in (AlCrNbSiTiV)N superhard coatings[J]. *Scripta Materialia*, 2010, 62, 105–108.

[36]   Shen WJ, Tsai MH, Tsai KY, et al. Superior Oxidation Resistance of $(Al_{0.34}Cr_{0.22}Nb_{0.11}Si_{0.11}Ti_{0.22})50N50$ High-Entropy Nitride. *Journal of the Electrochemical Society*, 2013, 160, C531.

[37]   Pogrebnjak AD, Yakushchenko IV, Bagdasaryan AA, et al. Microstructure, physical and chemical properties of nanostructured (Ti–Hf–Zr–V–Nb)N coatings under different deposition conditions[J]. *Materials Chemistry and Physics*, 2014, 147, 1079–1091.

[38]   Hsueh HT, Shen WJ, Tsai MH, et al. Effect of nitrogen content and substrate bias on mechanical and corrosion properties of high-entropy films (AlCrSiTiZr)100 – xNx. *Surface and Coatings Technology*, 2012, 206, 4106–4112.

[39]   Massiani Y, Medjahed A, Gravier P, et al. Electrochemical study of titanium nitride films obtained by reactive sputtering[J]. *Thin Solid Films*, 1990, 191, 305–316.

[40]   Massiani Y, Crousier J, Fedrizzi L, et al. Study of the behaviour in acidic solution of titanium and TiN coatings obtained by cathodic sputtering. *Surface and Coatings Technology*, 1987, 33, 309–317.

[41]   Milošev I, Strehblow HH, Navinšek B. Comparison of TiN, ZrN and CrN hard nitride coatings: Electrochemical and thermal oxidation[J]. *Thin Solid Films*, 1997, 303, 246–254.

[42]   Wiiala UK, Penttinen IM, Korhonen AS, et al. Improved corrosion resistance of physical vapour deposition coated TiN and ZrN. *Surface and Coatings Technology*, 1990, 41, 191–204.

[43]   Piippo J, Elsener B, Bohni H. Fatigue performance and tribological properties of r.f. sputtered TiN coatings. *Surface and Coatings Technology*, 1993, 61, 43–46.

[44]   von Fieandt K, Paschalidou EM, Srinath A, et al. Multi-component (Al,Cr,Nb,Y,Zr)N thin films by reactive magnetron sputter deposition for increased hardness and corrosion resistance. *Thin Solid Films*, 2020, 693, 137685.

[45]   Gild J, Braun J, Kaufmann K, et al. A high-entropy silicide: (Mo0.2Nb0.2Ta0.2Ti0.2W0.2)Si2. *Journal of Materiomics*, 2019, 5(3), 337–343. https://doi.org/10.1016/j.jmat.2019.03.002.

[46]   Qin Y, Liu JX, Li F, et al. A high entropy silicide by reactive spark plasma sintering. *Journal of Advanced Ceramics*, 2019, 8, 148–152. https://doi.org/10.1007/s40145-019-0319-3.

[47]   Lei C, Kai W, Wentao S, et al. Research progress of transition metal non-oxide high-entropy ceramics [J]. *Journal of Inorganic Materials*, 2020, 35(7), 748–758. https://doi.org/10.15541/jim20190408.

[48]   Hsieh MH, Tsai MH, Shen WJ, et al. Structure and properties of two Al-Cr-Nb-Si-Ti high-entropy nitride coatings[J]. *Surface & Coatings Technology*, 2013, 221, 118–123.

[49]   Wen T, Liu H, Ye B, et al. High-entropy alumino-silicides: A novel class of high-entropy ceramics. *Science China Materials*, 2020, 63, 300–306. https://doi.org/10.1007/s40843-019-9585-3.

[50] Dąbrowa J, Stygar M, Mikuła A, et al. Synthesis and microstructure of the (Co, Cr, Fe, Mn, Ni)3O4 high entropy oxide characterized by spinel structure. *Materials Letters*, 2018, 216, 32–36.

[51] Liu YC, Jia DC, Zhou Y, et al. Zn0.1Ca0.1Sr0.4Ba0.4ZrO3: A non-equimolar multi-component perovskite ceramic with low thermal conductivity. *Journal of the European Ceramic Society*, 2020, 40, 6272–6277.

[52] Chen H, Xiang HM, Dai FZ, et al. High entropy (Yb0.25Y0.25Lu0.25Er0.25)2SiO5 with strong anisotropy in thermal expansion. *Journal of Materials Science & Technology*, 2020, 36, 134–139.

[53] Zhang WM, Zhao B, Xiang HM, et al. One-step synthesis and electromagnetic absorption properties of high entropy rare earth hexaborides (HE REB6) and high entropy rare earth hexaborides/borates (HE REB6/HEREBO3) composite powders. *Journal of Advanced Ceramics*, 2021, 10, 62–77.

[54] Chen H, Zhao B, Zhao ZF, et al. Achieving strong microwave absorption capability and wide absorption bandwidth through a combination of high entropy rare earth silicide carbides/rare earth oxides. *Journal of Materials Science & Technology*, 2020, 47, 216–222.

[55] Castle E, Csanádi T, Grasso S, et al. Processing and properties of high-entropy ultra-high temperature carbides. *Scientific Reports*, 2018, 8, 8609.

[56] Zhang GR, Milisavljevic I, Zych E, et al. High-entropy sesquioxide X2O3 upconversion transparent ceramics. *Scripta Materialia*, 2020, 186, 19–23.

[57] Zhang RZ, Gucci F, Zhu HY, et al. Data-driven design of ecofriendly thermoelectric high-entropy sulfides. *Inorganic Chemistry*, 2018, 57, 13027–13033.

[58] Zhao ZF, Xiang HM, Dai FZ, et al. (La0.2Ce0.2Nd0.2Sm0.2Eu0.2)2Zr2O7: A novel high-entropy ceramic with low thermal conductivity and sluggish grain growth rate. *Journal of Materials Science & Technology*, 2019, 35, 2647–2651.

[59] Sarkar A, Loho C, Velasco L, et al. Multicomponent equiatomic rare earth oxides with a narrow band gap and associated praseodymium multivalency. *Dalton Transactions*, 2017, 46, 12167–12176.

[60] Chen H, Fu J, Zhang PF, et al. Entropy-stabilized metal oxide solid solutions as CO oxidation catalysts with high-temperature stability. *Journal of Materials Chemistry A*, 2018, 6, 11129–11133.

[61] Sarkar A, Velasco L, Wang D, et al. High entropy oxides for reversible energy storage. *Nature Communications*, 2018, 9, 3400.

[62] Mao AQ, Xiang HZ, Zhang ZG, et al. Solution combustion synthesis and magnetic property of rock-salt (Co0.2Cu0.2Mg0.2Ni0.2Zn0.2)O high-entropy oxide nanocrystalline powder. *Journal of Magnetism and Magnetic Materials*, 2019, 484, 245–252.

[63] Zhao ZF, Xiang HM, Chen H, et al. High-entropy (Nd0.2Sm0.2Eu0.2Y0.2Yb0.2)4A l2O9 with good high temperature stability, low thermal conductivity, and anisotropic thermal expansivity. *Journal of Advanced Ceramics*, 2020, 9, 595–605.

[64] Zhao ZF, Chen H, Xiang HM, et al. (La0.2Ce0.2Nd0.2Sm0.2Eu0.2)PO4: A high-entropy rare-earth phosphate monazite ceramic with low thermal conductivity and good compatibility with Al2O3. *Journal of Materials Science & Technology*, 2019, 35, 2892–2896.

[65] Lei ZF, Liu XJ, Li R, et al. Ultrastable metal oxide nanotube arrays achieved by entropy-stabilization engineering. *Scripta Materialia*, 2018, 146, 340–343.

[66] Zhang JR, Zhang XY, Li Y, et al. High-entropy oxides 10La2O3–20TiO2–10Nb2O5–20WO3–20ZrO2 amorphous spheres prepared by containerless solidification. *Materials Letters*, 2019, 244, 167–170.

[67] Guo YC, Li JQ. Preparation of high-entropy (ReTiZrYAl)O glasses by aerodynamic levitation and performance study. *Proceedings of the 21st National Annual Conference on High Technology Ceramics*, 2020, 15–10.

[68]   Chen XQ, Wu YQ. High-entropy transparent fluoride laser ceramics. *Journal of the American Ceramic Society*, 2020, 103, 750–756.

[69]   Zhao ZF, Chen H, Xiang HM, et al. High entropy defective fluorite structured rare-earth niobates and tantalates for thermal barrier applications. *Journal of Advanced Ceramics*, 2020, 9, 303–311.

[70]   Okejiri F, Zhang ZH, Liu JX, et al. Room-temperature synthesis of high-entropy perovskite oxide nanoparticle catalysts through ultrasonication-based method. *ChemSusChem*, 2020, 13, 111–115.

[71]   Wang KW, Ma BS, Li T, et al. Fabrication of high-entropy perovskite oxide by reactive flash sintering. *Eram International*, 2020, 46, 18358–18361.

[72]   Dąbrowa J, Stygar M, Mikuła A, et al. Synthesis and microstructure of the (Co, Cr, Fe, Mn, Ni)3O4 high entropy oxide characterized by spinel structure. *Materials Letters*, 2018, 216, 32–36.

[73]   Pu YP, Zhang QW, Li R, et al. Dielectric properties and electrocaloric effect of high-entropy (Na0.2Bi0.2Ba0.2Sr0.2Ca0.2)TiO3 ceramic. *Applied Physics Letters*, 2019, 115, 223901.

[74]   Dong Y, Ren K, Lu YH, et al. High-entropy environmental barrier coating for the ceramic matrix composites. *Journal of the European Ceramic Society*, 2019, 39, 2574–2579.

[75]   Moskovskikh DO, Vorotilo S, Sedegov AS, et al. high-entropy (HfTaTiNbZr)C and (HfTaTiNbMo)C carbides fabricated through reactive high-energy ball milling and spark plasma sintering. *Ceramics International*, 2020, 46, 9008–19014.

[76]   Guan JY, Li DX, Yang ZH, et al. Synthesis and thermal stability of novel high-entropy metal boron carbonitride ceramic powders. *Ceramics International*, 2020, 46, 26581–26589.

[77]   Ning SS, Wen TQ, Ye BL, et al. Low-temperature molten salt synthesis of high-entropy carbide nanopowders. *Journal of the American Ceramic Society*, 2020, 103, 2244–2251.

[78]   Feng L, Fahrenholtz WG, Hilmas GE, et al. Synthesis of ingle-phase high-entropy carbide powders. *Scripta Materialia*, 2019, 162, 90–93.

[79]   Liu D, Wen TQ, Ye BL, et al. Synthesis of superfine high-entropy metal diboride powders. *Scripta Materialia*, 2019, 167, 110–114.

[80]   Zhang Y, Jiang ZB, Sun SK, et al. Microstructure and mechanical properties of high-entropy borides derived from boro/carbothermal reduction. *Journal of the European Ceramic Society*, 2019, 39, 3920–3924.

[81]   Feng L, Fahrenholtz WG, Hilmas GE. Low-temperature sintering of single-phase, high-entropy carbide ceramics. *Journal of the American Ceramic Society*, 2019, 102, 7217–7224.

[82]   Zhang Y, Sun SK, Zhang W, et al. Improved densification and hardness of high-entropy diboride ceramics from fine powders synthesized via borothermal reduction process. *Ceramics International*, 2020, 46, 14299–14303.

[83]   Gild J, Zhang Y, Harrington T, et al. High-entropy metal diborides: A new class of high-entropy materials and a new type of ultrahigh temperature ceramics. *Scientific Reports*, 2016, 6, 37946.

[84]   Zhang Y, Guo WM, Jiang ZB, et al. Dense high-entropy boride ceramics with ultra-high hardness. *Scripta Materialia*, 2019, 164, 135–139.

[85]   Gu JF, Zou J, Sun SK, et al. Dense and pure high-entropy metal diboride ceramics sintered from self-synthesized powders via boro/carbothermal reduction approach. *Science China Materials*, 2019, 62, 1898–1909.

[86] Chen H, Zhao ZF, Xiang HM, et al. Effect of reaction routes on the porosity and permeability of porous high entropy (Y0.2Yb0.2Sm0.2Nd0.2Eu0.2)B6 for transpiration cooling. *Journal of Materials Science & Technology*, 2020, 38, 80–85.

[87] Feng L, Fahrenholtz WG, Hilmas GE. Processing of dense high-entropy boride ceramics. *Journal of the European Ceramic Society*, 2020, 40, 3815–3823.

[88] Castle E, Csanádi T, Grasso S, et al. Processing and properties of high-entropy ultrahigh temperature carbides. *Scientific Reports*, 2018, 8, 8609.

[89] Sarker P, Harrington T, Toher C, et al. High-entropy high-hardness metal carbides discovered by entropy descriptors. *Nature Communications*, 2018, 9, 4980.

[90] Chen H, Xiang HM, Dai FZ, et al. Low thermal conductivity and high porosity ZrC and HfC ceramics prepared by in situ reduction reaction/partial sintering method for ultrahigh temperature applications. *Journal of Materials Science & Technology*, 2019, 35, 2778–2784.

[91] Chen H, Xiang HM, Dai FZ, et al. High strength and high porosity YB2C2 ceramics prepared by a new high temperature reaction/partial sintering process. *Journal of Materials Science & Technology*, 2019, 35, 2883–2891.

[92] Guo QQ, Xiang HM, Sun X, et al. Preparation of porous YB4 ceramics using a combination of in situ borothermal reaction and high temperature partial sintering. *Journal of the European Ceramic Society*, 2015, 35, 3411–3418.

[93] Chen H, Xiang HM, Dai FZ, et al. High porosity and low thermal conductivity high entropy (Zr0.2Hf0.2Ti0.2Nb0.2Ta0.2)C. *Journal of Materials Science & Technology*, 2019, 35, 1700–1705.

[94] Liu RH, Chen HY, Zhao KP, et al. Entropy as a genelike performance indicator promoting thermoelectric materials. *Advanced Materials*, 2017, 29, 1702712.

[95] Meisenheimer PB, Kratofil TJ, Heron JT. Giant enhancement of exchange coupling in entropy-stabilized oxide heterostructures. *Scientific Reports*, 2017, 7, 13344.

[96] Kotsonis GN, Rost CM, Harris DT, et al. Epitaxial entropy-stabilized oxides: Growth of chemically diverse phases via kinetic bombardment. *MRS Communications*, 2018, 8, 1371–1377.

[97] Yeh JW, Chen SK, Lin SJ, et al. Nanostructured high-entropy alloys with multiple principal elements: Novel alloy design concepts and outcomes. *Advanced Engineering Materials*, 2004, 6, 299–303.

[98] Cantor B, Chang ITH, Knight P, et al. Microstructural development in equiatomic multicomponent alloys. *Materials Science and Engineering: A*, 2004, 375, 213–218.

[99] Wang X, Zhang Y, Qiao Y, et al. Novel microstructure and properties of multicomponent CoCrCuFeNiTix alloys. *Intermetallics*, 2007, 15, 357–362.

[100] Zhang Y, Zhou Y, Lin J, et al. Solid-solution phase formation rules for multi-component alloys. *Advanced Engineering Materials*, 2008, 10, 534–538.

[101] Li C, Li JC, Zhao M. Effect of alloying elements on microstructure and properties of multiprincipal elements high-entropy alloys[J]. *Journal of Alloys and Compounds: An Interdisciplinary Journal of Materials Science and Solid-State Chemistry and Physics*, 2009, 475(1–2), 752–757.

[102] Senkov O, Wilks G, Miracle D, et al. Refractory high-entropy alloys[J]. *Intermetallics*, 2010, 18(9), 1758–1765.

[103] Singh S, Wanderka N, Murty BS, et al. Decomposition in multi-component AlCoCrCuFeNi high-entropy alloy[J]. *Acta Materialia*, 2011, 59(1), 182–190

[104] Senkov ON, Wilks GB, Scott JM, et al. Mechanical properties of NbMoTaW and VNbMoTaW refractory high entropy alloys[J]. *Intermetallics*, 2011, 19(5), 698–706.

[105] Lin C-M, Tsai H-L. Evolution of microstructure, hardness, and corrosion properties of high-entropy Al0.5CoCrFeNi alloy. *Intermetallics*, 2011, 19, 288–294.

[106] Yang X, Zhang Y, Liaw P. Microstructure and compressive properties of NbTiVTaAlx high entropy alloys. *Procedia Engineering*, 2012, 36, 292–298.

[107] Yang X, Zhang Y. Prediction of high-entropy stabilized solid-solution in multi-component alloys. *Materials Chemistry and Physics*, 2012, 132, 233–238.

[108] Zhang Y, Zuo T, Cheng Y, et al. High-entropy alloys with high saturation magnetization, electrical resistivity and malleability. *Scientific Reports*, 2013, 3, 1455. https://doi.org/10.1038/srep01455.

[109] Yeh JW, Lin SJ, Chin TS, et al. Formation of simple crystal structures in Cu-Co-Ni-Cr-Al-Fe-Ti-V alloys with multiprincipal metallic elements. *Metallurgical and Materials Transactions A*, 2004, 35, 2533–2536.

[110] Cantor B, Audebert F, Galano M, et al. Novel multicomponent alloys. *Journal of Metastable and Nanocrystalline Materials*, 2005, 24–25, 1–6.

[111] Zhang Y, Lu ZP, Ma SG, et al. Guidelines in predicting phase formation of high-entropy alloys. *MRS Communications*, 2014, 4, 57–62. https://doi.org/10.1557/mrc.2014.11.

[112] Ma SG, Liaw PK, Gao MC, et al. Damping behavior of AlxCoCrFeNi high-entropy alloys by a dynamic mechanical analyzer. *Journal of Alloys and Compounds*, 2014, 604, 331–339.

[113] Chang HW, Huang PK, Davison A, et al. Nitride films deposited from an equimolar Al-Cr-Mo-Si-Ti alloy target by reactive direct current magnetron sputtering. *Thin Solid Films*, 2008, 516, 6402–6408.

[114] Zhang Y. Mechanical properties and structures of high entropy alloys and bulk metallic glasses composites. *Materials Science Forum*, 2010, 654–656, 1058–1061.

[115] Zhang H, Pan Y, He Y, Jiao H. Microstructure and properties of 6FeNiCoSiCrAlTi high-entropy alloy coating prepared by laser cladding. *Applied Surface Science*, 2011, 257, 2259–2263.

[116] Lin C-M, Tsai H-L, Bor H-Y. Effect of aging treatment on microstructure and properties of high-entropy Cu0.5CoCrFeNi alloy. *Intermetallics*, 2010, 18(6), 1244–1250.

[117] Xiang H, Xing Y, Dai F, et al. High-entropy ceramics: Present status, challenges, and a look forward. *Journal of Advanced Ceramics*, 2021, 10, 385–441. https://doi.org/10.1007/s40145-021-0477-y.

[118] Chen K, Pei X, Tang L, et al. A five-component entropy-stabilized fluorite oxide [J]. *Journal of the European Ceramic Society*, 2018, 38(11), 4161–4164.

[119] Gild J, Samiee M, Braun JL, et al. High-entropy fluorite oxides [J]. *Journal of the European Ceramic Society*, 2018, 38(10), 3578–3584.

[120] Liu R, Chen H, Zhao K, et al. Entropy as a gene like performance indicator promoting thermoelectric materials [J]. *Advanced Materials*, 2017, 29(38), 1702712.

[121] Huang Z, Miller SA, Ge B, et al. High thermoelectric performance of new rhombohedral phase of GeSe stabilized through alloying with AgSbSe2 [J]. *Angewandte Chemie International Edition*, 2017, 56(45), 14113–14118.

[122] Roychowdhury S, Ghosh T, Arora R, et al. Stabilizing n-type cubic GeSe by entropy-driven alloying of AgBiSe2: Ultralow thermal conductivity and promising thermoelectric performance [J]. *Angewandte Chemie-International Edition*, 2018, 57(46), 15167–15171.

[123] Braic V, Vladescu A, Balaceanu M, Luculescu CR, Braic M. Nanostructured multi-element (TiZrNbHfTa)N and (TiZrNbHfTa)C hard coatings. *Surface and Coatings Technology*, 2012, 211, 117–121.

[124] Chen, T-K, Wong M-S. Structure and properties of reactively-sputtered AlxCoCrCuFeNi oxide films. *Thin Solid Films*, 2007, 516, 141–146.

[125] Jiang B, Yu Y, Cui J, et al. Hunt for renewable plastics clears a hurdle. *Science*, 2021, 371, 830.

[126] Lin M-I, Tsai M-H, Shen W-J, et al. Evolution of structure and properties of multi-component (AlCrTaTiZr)Ox films. *Thin Solid Films*, 2010, 518, 2732–2737.

[127] Zheng Y, Zou M, Zhang W, et al. Electrical and thermal transport behaviours of high-entropy perovskite thermoelectric oxides. *Journal of Advanced Ceramics*, 2021, 10, 377–384. https://doi.org/10.1007/s40145-021-0462-5.

[128] Jiang B, Yu Y, Chen H, et al. Entropy engineering promotes thermoelectric performance in p-type chalcogenides. *Nature Communications*, 2021, 12, 3234. https://doi.org/10.1038/s41467-021-23569-z.

[129] Albedwawi SH, AlJaberi A, Haidemenopoulos GN, et al. High entropy oxides-exploring a paradigm of promising catalysts: A review. *Materials & Design*, 2021, 202, 109534. https://doi.org/10.1016/j.matdes.2021.109534.

[130] George EP, Raabe D, Ritchie RO. High-entropy alloys. *Nature Reviews Materials*, 2019, 4, 515–534.

[131] Zhang G, Ming K, Kang J, et al. High entropy alloy as a highly active and stable electrocatalyst for hydrogen evolution reaction. *Electrochimica Acta*, 2018, 279, 19–23.

[132] Wang A-L, Wan H-C, Xu H et al. Quinary PdNiCoCuFe alloy nanotube arrays as efficient electrocatalysts for methanol oxidation. *Electrochimica Acta*, 2014, 127, 448–453.

[133] Yusenko KV, Riva S, Carvalho PA, et al. First hexagonal close packed high-entropy alloy with outstanding stability under extreme conditions and electrocatalytic activity for methanol oxidation. *Scripta Materialia*, 2017, 138, 22–27.

[134] Qiu H-J, Fang G, Wen Y, et al. Nanoporous high-entropy alloys for highly stable and efficient catalysts. *Journal of Materials Chemistry A*, 2019, 7, 6499–6506.

[135] Yao Y, Huang Z, Xie P, et al. Carbothermal shock synthesis of high-entropy-alloy nanoparticles. *Science*, 2018, 359, 1489–1494.

[136] Cui X, Zhang B, Zeng C, et al. Electrocatalytic activity of high-entropy alloys toward oxygen evolution reaction. *MRS Communications*, 2018, 8, 1230–1235.

[137] Xie P, Yao Y, Huang Z, et al. Highly efficient decomposition of ammonia using high-entropy alloy catalysts. *Nature Communications*, 2019, 10, 4011.

[138] Löffler T, Meyer H, Savan A, et al. Discovery of a multinary noble metal-free oxygen reduction catalyst. *Advanced Energy Materials*, 2018, 8, 1802269.

[139] Batchelor TAA, Pedersen JK, Winther SH, et al. High-entropy alloys as a discovery platform for electrocatalysis. *Joule*, 2019, 3, 834–845.

[140] Lv ZY, Liu XJ, Jia B, et al. Development of a novel high-entropy alloy with eminent efficiency of degrading azo dye solutions. *Scientific Reports*, 2016, 6, 34213.

[141] Chen H, Fu J, Zhang P, et al. Entropy-stabilized metal oxide solid solutions as CO oxidation catalysts with high-temperature stability. *Journal of Materials Chemistry A*, 2018, 6, 11129–11133.

[142] Chen H, Lin W, Zhang Z, et al. Mechanochemical synthesis of high entropy oxide materials under ambient conditions: Dispersion of catalysts via entropy maximization. *ACS Materials Letters*, 2019, 1, 83–88.

[143] Chen H, Lin W, Zhang Z, et al. Mechanochemical synthesis of high entropy oxide materials under ambient conditions: Dispersion of catalysts via entropy maximization. *ACS Materials Letters*, 2019, 1, 83–88.

[144] Gong K, Du F, Xia Z, et al. Nitrogen-doped carbon nanotube arrays with high electro-catalytic activity for oxygen reduction. *Science*, 2009, 323, 760–764.

[145] Lefèvre M, Proietti E, Jaouen F, et al. Iron-based catalysts with improved oxygen reduction activity in polymer electrolyte fuel cells. *Science*, 2009, 324, 71–74.

[146] Jaouen F, Proietti E, Lefèvre M, et al. Recent advances in non-precious metal catalysis for oxygen-reduction reaction in polymer electrolyte fuel cells. *Energy & Environmental Science*, 2011, 4, 114–130.

[147] Jaouen F, Proietti E, Lefèvre M, et al. Recent advances in non-precious metal catalysis for oxygen-reduction reaction in polymer electrolyte fuel cells. *Energy & Environmental Science*, 2011, 4, 114–130.

[148] Li Y, Zhou W, Wang H, et al. An oxygen reduction electrocatalyst based on carbon nanotube—graphene complexes. *Nature Nanotechnology*, 2012, 7, 394–400.

[149] Banham D, Ye S, Pei K, et al. A review of the stability and durability of non-precious metal catalysts for the oxygen reduction reaction in proton exchange membrane fuel cells. *Journal of Power Sources*, 2015, 285, 334–348.

[150] Qin Y, Liu JX, Li F, et al. A high entropy silicide by reactive spark plasma sintering. *Journal of Advanced Ceramics*, 2019, 8, 148–152.

[151] Zhou Y, Zhang Y, Wang Y, et al. Solid solution alloys of AlCoCrFeNiTix with excellent room-temperature mechanical properties. *Applied Physics Letters*, 2007, 90, 181904.

[152] Zhang Y, Zhou YJ. Mater solid solution formation criteria for high entropy alloys. *Materials Science Forum*, 2007, 561–565, 1337–1339.

[153] Cai H, Zhang H, Zheng Y. Soft magnetic devices applied for low zero excursion (0.01-/h) four-mode ring laser gyro. *IEEE Transactions on Magnetics*, 2007, 43(6), 2686–2688. http://doi.org/10.1109/TMAG.2007.893315.

[154] Lu Z, Chen G, Siahrostami S, et al. High-efficiency oxygen reduction to hydrogen peroxide catalysed by oxidized carbon materials. *Nature Catalysis*, 2018, 1, 156–162.

9 High-Entropy Polymers and Entropic Materials

Yong Zhang and Yuanying Yue

9.1 INTRODUCTION

With the development of science and the progress of society, the demand for high-performance materials is increasing. Materials made up of a single element can no longer meet people's needs, and new materials are urgently needed. High-entropy materials, a class of materials with a single phase obtained by a mutual solid solution with equimolar or near-molar ratios containing five or more elements, have been proposed in recent years, and their unique phase structure and functional tunability have attracted wide attention from researchers [1,2]. In 2004, Yeh et al. [3] proposed that the increase in mixed-configuration entropy caused by the increase in constituent elements is sufficient to overcome the enthalpy of the formation of single-phase compounds and, thus, inhibit the formation of brittle intermetallic compounds and defined for the first time an alloy composed of five or more elements in equal proportions based on the magnitude of mixed-configuration entropy as a "high-entropy" alloy. The alloy composed of five or more elements in equal proportions was defined as a "high-entropy alloy" for the first time. Since the thermodynamic stability is determined by the minimization of Gibbs free energy ($G = H - TS$, where G is Gibbs free energy, H is enthalpy, S is entropy, and T is temperature), high-entropy materials with lower Gibbs free energy can be more stable at high temperatures. As the research progressed, researchers found that high-entropy alloys exhibit unique effects in thermodynamics, kinetics, microstructure, and properties; namely, the high-entropy effect, sluggish diffusion effect, severe-lattice-distortion effect, and cocktail effect [4–6]. Compared with traditional alloys, high-entropy alloys have excellent mechanical properties, thermodynamic stability, electrical and magnetic properties, and catalytic activity, and are expected to be used in high-temperature, wear- and corrosion-resistant scenarios, as well as energy and environmental fields [7–11].

The success of high-entropy alloys has inspired researchers to explore other types of high-entropy materials. Recent advances in high-entropy technologies include high-entropy ceramics (HECs), high-entropy polymers (HEPs), and high-entropy composites (HECOMPs) [12]. HEAs, however, have only been adapted in part by a few polymers. Polymers borrowed from the concept of high-entropy alloys can enhance their performance in two ways: first, by blending multiple

DOI: 10.1201/9781003319986-9

polymers, where the entropy is beneficial in reducing polymer phase separation, and second, by compounding high-entropy alloys with polymers to prepare high-entropy composites that can complement and strengthen the material in terms of performance [13,14]. It is well known that polymers are long-chain molecules, and usually, the mixing entropy decreases with the size of the blending entity; therefore, high-entropy polymers are rarely studied because the vast majority of polymers are incompatible with each other since the blending performance of polymers not only depends on the polymer itself but is also closely related to the phase morphology of the internal components [15]. Poor mixing not only does not improve the performance of the material but also significantly reduces the performance of the material because of the reduced adhesion at the phase interface. Polymer blends are generally phase separated and exhibit poor performance as well as instability. Most polymer blends exhibit micron-scale phase separation, including matrix dispersion, matrix fibers, to lamellar structures. Such microphase interfaces often become the origin of cracks and catastrophic failures, limiting their further application. The conformational entropy of high-entropy alloys can suppress the appearance of intermetallic compounds [8], the mixing entropy facilitates polymer phase mixing, and the contribution of entropy favors random mixing, chain diffusion, and entanglement to maximize randomness and disorder in the system and reduce the possibility of phase separation. The exploration of polymer blends and the study of polymer composites hold great promise for the exploration of higher-efficiency photovoltaic properties, irradiation resistance, biomedicine, and polymers for batteries with a large storage capacity [12,16,17]. This chapter will summarize the current development and applications of high-entropy polymers and present the views on the future trends of high-entropy polymers.

9.2 HIGH-ENTROPY POLYMER

A polymer is a long-chain molecule or macromolecule that is built by connecting repeating chemical units. Each molecule of a polymer can be composed of hundreds, thousands, or even millions of repeating units. Polymers have promising applications in energy, medical, and military fields [18,19]. The concept of high entropy is not limited to metallic materials, nor is it defined only by the number of elements and the disorder of their composition. The fact that the combination of multiple elements can be achieved in HEAs has inspired new design strategies in the field of organic materials. Organic materials in high-entropy configurations, such as polymers and carbon materials, can be considered an effective way to design interesting new materials [20]. In contrast to alloys, the space-filling scheme of the structural unit of the polymer, the molecular motif, is mainly determined by lattice interactions. Therefore, the entropy of organic eutectic comes mainly from the acceptable change of molecular conformation. Different types and proportions of repeating units of polymers increase the lattice stacking entropy and affect the phase transition of crystalline polymers. In this respect, entropy is described as the degree of disorder in a system. According to the Flory-Huggins (FH) theory, the polymer solution system is considered a lattice-like system [21], and the influence of the difference in the size of polymer and solvent molecules on the change of mixing entropy is considered. The entropy of mixing two molecules is inversely proportional to the degree of polymerization (1/N) [22]. The development of new high-entropy polymers drawing on the definition of high-entropy alloys can start with the blending of multiple polymers. Polymer blending can be divided into two methods, physical and chemical blending [23].

9.2.1 Physical Blending

Physical blending only refers to the shear, diffusion, convection, stretching, and other physical effects to achieve the purpose of a mixing method. It does not involve any chemical reaction. This method is simple to operate, but it is often difficult to achieve a good dispersion effect, as it easily appears in the subsequent process of delamination, it disperses together, its size increases, and an embrittlement phenomenon occurs, which is not conducive to the improvement of the performance of the blend. Adding graft or block copolymer as compatibilizers can obviously improve this phenomenon. The main chain and branch chain or joint chain of the copolymer can be compatible with different blends, respectively, so they can be used as surfactants to reduce the interfacial tension, improve the binding force between the two phases, and improve the physical properties of the blends.

9.2.2 Chemical Blending

Chemical blending is a compatible method for components going through chemical reactions to make a polymer, including grafting blending, block blending, interpenetrating polymer network (IPN) blending technology, and the methods of reactive extrusion technology.

Graft blending is the first chemical blending method to obtain a large-scale industrial application. The active center of the second monomer can be generated on the main chain of polymerization, and then the branch chain can be formed. The active functional groups distributed on the main chain can also be used to conjugate with another polymer with reactive groups at the end of the molecule to form branched polymer compounds. ABS and HIPS are typical examples of graft blending. Block blending is the polymerization of polymer chain ends of each component, or the linear macromolecules with polymerizable functional groups participate in other monomer polymerization reactions to form block copolymers. By adjusting the number and length of each block, a series of products with adjustable physical properties can be obtained. The IPN blending alloy is a hybrid system in which two or more kinds of polymer chains interpenetrate and intertwine with each other and has a microphase separation structure formed by a cross-linking network. Cross-linking can be chemical or physical. Therefore, strictly speaking, IPN technology belongs to the physicochemical blending method. IPN technology enables the two polymers to penetrate each other, the two phases have good dispersion, the phase interface is large, and it can play a good synergistic effect. Reactive extrusion blending is a method of blending in a twin-screw extruder with simultaneous chemical reactions. The twin-screw extruders used in this method are mostly meshing and rotating in the same direction, which requires the material supply and transportation to be continuous and stable, the adjustment of reaction temperature and time to be accurate and convenient, the screw length and diameter to be relatively large, and the material residence time to be long.

The compatibility of two polymers depends first on the structure of the polymer itself [24]. Thermodynamically, the compatibility is determined by the free energy of mixing (ΔG). That is, the mixed system can be compatible only when $\Delta G < 0$. Thermodynamic compatibility refers to the ability to form a homogeneous system at any ratio. For polymers, the mixing entropy is generally small due to the large molecular weights, so it is the mixing enthalpy that mostly determines the effect. However, with the emergence of high-entropy alloys, the concern is that mixed entropy also plays a huge role in it. In addition to the thermodynamic compatibility described earlier, polymer alloys also have process compatibility with the two polymers [25]. Process compatibility refers to the ability of two polymers to readily disperse with each other to produce a stable blend. This process compatibility is determined by the kinetic factors involved in the blending. For example, two polymers, despite their close solubility parameters and good thermodynamic compatibility, still cannot be miscible due to excessive molecular weight, high crystallinity, high viscosity, and unsuitable mixing conditions. Conversely, when two polymers are poorly compatible and are mixed by mechanical methods or other conditions of appropriate degree, it is possible to obtain a sufficiently stable blend product. This is due to the exceptionally high viscosity of the polymer and the difficulty in moving the molecular chain segments. Although there is a thermodynamic tendency for automatic separation into two phases, in practice, the phase separation is extremely slow, so it is difficult to separate the blended system into two macroscopic phases over a very long period. However, due to the great differences in molecular structure, polarity, solubility parameters, and molecular weight of polymers, as well as the high viscosity caused by high molecular weight, few alloy systems can achieve sufficient compatibility even under strong mechanical action. Currently, most important blends are compatible only to the extent of partial slight mixing. To obtain more polymer-blended alloy materials of practical value, effective measures must be taken to capacitate the polymer blends. High-entropy effects play an important role in polymer blending.

9.3   HIGH-ENTROPY CONCEPT IS EXTENDED TO HEPs

The direct application of HEAs design concepts to polymers has attracted researchers' attention in recent years. In high-entropy alloys, when the effect of configuration entropy is greater than the effect of enthalpy, the alloy forms a single solid-solution phase, such as FCC, BCC, and HCP, with high strength, corrosion resistance, wear resistance, high temperature resistance, and other properties. The application of the concept of high-entropy to polymer blending has led to new discoveries. Wu et al. [26] induced co-crystallization of dissolved polyvinylidene fluoride (PVDF) homopolymer and polyvinylidene fluoride (PVDF-TrFE) copolymer solutes by solvent evaporation. The role of lattice stacking entropy on the Curie jump of eutectic is revealed, as well as the change of stacking dispersion of wafer layers inside the material related to the change of the base sequence composition is elaborated. In the mixing of polymers, the synergistic effect of strong enthalpy tends to overlook the role of weak entropy, but the role of entropy leads to the discovery of different possibilities. Huo et al. [27] investigated that entropy in binary blends plays an important role in facilitating the realization of nanostructured materials. It was demonstrated that the binary polylactic acid (PLA)/polymethyl methacrylate (PMMA) blend formed a super-tough polymer with a strength of about 70 MPa and a tensile toughness of about 60 MJ/m^3. PLA and PMMA are brittle at room temperature, with elongation at break of ~3% and ~9%, respectively. At the polymer chain level, entropy drives the mixing of the different chains, so PLA-PMMA exhibits a higher entanglement density above linearity, resulting in a significant increase in elongation at break of PLA-75 up to ~146% strain, which is a 50-fold increase compared to pure PLA. The effect of entropy in binary polymer blends can contribute to the formation of high-tenacity polymers, but the effect of entropy in ternary polymer systems or multiple systems remains to be investigated. Huang et al. [14] blended up to five ($n = 5$) different polymers of polystyrene (PS), poly(methyl methacrylate) (PMMA), polycarbonate (PC), polyvinylpyrrolidone (PVP), and polyisoprene (PIP) into solid films by spin-coating on slides using a common solvent (Figure 9.1).

It was found that the presence of many different polymer segments reduced the probability of meeting and hindered polymerization, thus inhibiting phase separation. The results show that the steric hindrance is stronger with higher molecular weight. With further large increases in n, the inhibition of unmixing of polymer blends may eventually make a significant contribution from the reduction in the free energy of mixing due to the additional entropy provided by the addition of species. The macroscopic phase separation is inhibited by increasing the entropy by increasing the types of polymers, but the effect of high entropy on atomic scale has not been studied. To investigate the high-entropy effect at the atomic scale, researchers prepared polymer blends using in situ exchange reactions. Hirai et al. [28] prepared binary to quaternary polymer blends using four aliphatic polyamides (PA6, PA610, PA11, and PA12) as component polymers. The effects of fractional and exchange reaction ratios on the crystallization behavior and mechanical properties of polymer blends were investigated, and the

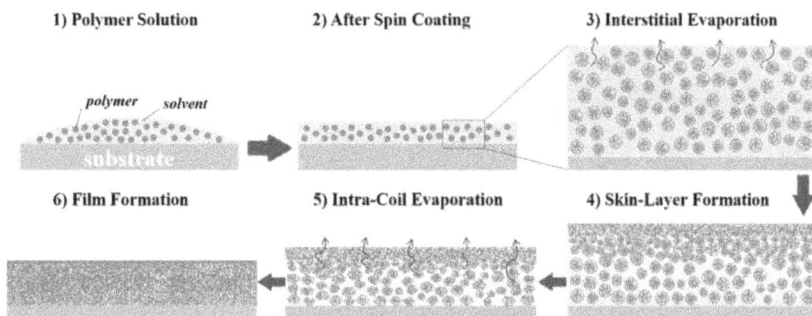

FIGURE 9.1 A schematic representation of the film formation process during spin-coating [14].

specific conditions for obtaining high-entropy polymer blends with specific properties were discussed. As shown in Figure 9.2, binary blends are divided into active blends (PA6/PA11, PA610/PA11, PA11/PA12) and low active blends (PA6/PA610, PA6/PA12, PA610/PA12). Three properties (crystallization temperature, elongation at break, and haze) were compared with the exchange reaction ratio. In general, polymer blends with a high-exchange reaction ratio have lower crystallization temperatures and higher elongation at break than blends with a low-exchange reaction ratio. Compared with binary blends with the same exchange reaction ratio, ternary blends and quaternary blends have lower T_c and higher elongation at break. Although the relationship between haze and reaction ratio is unclear, the multicomponent blends after the reaction often exhibits excellent transparency. High entropy in polymer blends can limit crystallinity and exhibit high elongation at break.

Although high-entropy polymers are still in the early stages of development, further studies can be conducted to explore new functional solid polymer electrolytes with enhanced ionic conductivity and cycling stability for Li/Na-ion batteries. Zhang et al. [16] demonstrated that low-enthalpy and high-entropy (LEHE) electrolytes can intrinsically produce very free ions and high mobility, allowing them to efficiently drive Li-ion storage. This LEHE electrolyte was constructed based on the introduction of $CsPbI_3$ calcium titanite quantum dots (PQDs) to enhance PEO@LiTF SI complexes. Thermodynamically, the LEHE effect promotes the degree of structural disorder of PEO, which promotes the dissociation of lithium salts and the generation of more free lithium ions. On the power side, it facilitates rapid ion transfer, uniform charge distribution, and free lithium dendrites. Ion-conducting solid polymer electrolytes have a promising future in the development of advanced lithium batteries,

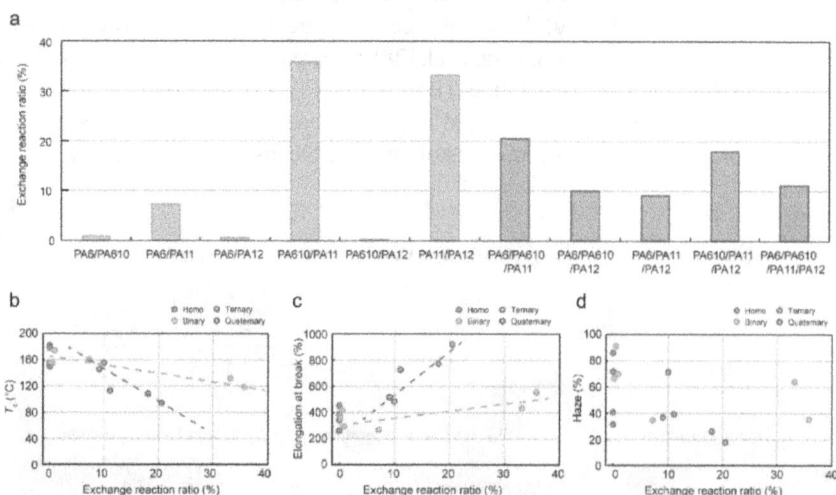

FIGURE 9.2 Evaluation of exchange reaction ratio and relationships between properties and exchange reaction ratio: (a) exchange reaction ratios of polymer blends; (b–d) relationships between exchange reaction ratio and (b) crystallization temperature (T_c), (c) elongation at break, and (d) haze [28].

improving the performance of lithium batteries by developing new high-entropy electrolytes and also simultaneously gaining a deeper understanding of ion transfer mechanisms. Moreover, many new high-entropy polymer-composite materials have the potential to be designed and studied in the future.

9.4   HIGH-ENTROPY COMPOSITION MATERIALS

In addition to directly applying the design concept of HEAs to polymers, another method to optimize polymers is to combine the HEAs with polymers into composition materials, such as metal-organic frameworks. The combination of magnetron sputtering and two-photon lithography to combine high-entropy alloys with polymer lattice structures is expected to challenge properties that cannot be achieved by conventional mechanics, providing a solution for new construction materials as well as flexible materials, such as those with high strength but low mass. For the first time, Gao et al. [29] used two-photon lithography combined with physical vapor deposition to prepare 80 nm-thick high-entropy alloy films ($CoCrFeNiAl_{0.3}$) clad with three-dimensional polymer nanolattices with feature sizes ranging from 5 nm to 20 μm. Combining the advantages of HEA and polymeric nanolattice structures, the hybridized nanolattice has superior compressive specific strength of 0.032 MPa $kg^{-1}m^3$ at densities below 1000 kg/m^3, while still providing good compressive resistance compared to pure metals/alloys. However, the composite nanolattice recovers at a strain of nearly 8%. This indicates that three-dimensional nanostructured metamaterials have unique mechanical properties, but there is a conflict between strength and recoverability that limits their applications in structural components, energy absorption, storage, and others. Recoverability represents the ability of a material to return to its initial state after being subjected to a large, applied strain in a plastic or inelastic state. The trade-off between strength and recoverability is precisely related to the conflict between strength and plasticity in materials, a competing mechanism between strengthening and toughening that has long existed in the mechanics of materials. To overcome the conflict between strength and recoverability, Zhang et al. [13] concluded that the dot-matrix material consisting of a single native material is an important reason why the conflict between strength and recoverability cannot be overcome, so a three-dimensional nanolattice was prepared by two-photon lithography and deposited by magnetron sputtering with composite nanostructures of high-entropy alloy films with thicknesses ranging from tens to hundreds of nanometers. In this case, the lightweight and ductile polymer core acts as a frame, allowing the entire structure to recover after large deformations, while the ultrastrong HEA coating increases strength. The composite nanolattice not only has a high specific strength but is also able to recover almost completely after compression at strains exceeding 50%. Through previous studies, it has been shown that composite materials of high-entropy alloys and three-dimensional nanostructured polymers have the potential to break through the limits of conventional mechanics and have promising applications. In addition, some scholars have also prepared composite materials by encapsulating high-entropy alloys in a polymer matrix to develop flexible materials with excellent radiation shielding properties, which are expected to be used in nuclear reactors, industry, radiotherapy, aerospace, and other fields. Wang et al. [30]

prepared soft elastic composites using GaInSnPbBi low melting point high-entropy alloy as filler with polydimethylsiloxane (PDMS). The composite material has the characteristics of light weight, good elasticity, and excellent radiation resistance.

The complexity of preparing composites from high-entropy alloys and polymers has led to few research so far; however, combining two materials is expected to break through the limitations of existing materials, such as the conflict of strength and recoverability, overcoming the limitations brought by a single material.

9.5 SUMMARY

Since the concept of high-entropy alloys was first introduced in 2004, high-entropy polymers have been developed and have become one of the major concerns in the research of special materials. One existing study directly follows the concept of high-entropy alloys to polymers, but the four main properties of high-entropy alloys have not been fully discovered in high-entropy polymers due to the difference in the base elements. Another is the preparation of composite materials by combining high-entropy alloys with polymers, which is expected to break through the conflict of strength and recoverability, and others, providing ideas for the development of new types of materials. The investigation of the properties and applications of high-entropy polymers is just beginning, and different preparation methods and processes, as well as their optimization, will have an impact on their performance. Overall, the concept of high-entropy polymers could pave the way for the development of advanced polymeric materials in the future. It is predicted that the main development directions of high-entropy alloy polymers in the future are as follows: (1) the study of new systems of high-entropy polymers; (2) the exploration and discovery of the properties of existing high-entropy polymers; (3) combination of theoretical calculations and experimental studies to investigate the composition and calculations of new high-entropy polymers, use experiments to achieve a purposeful synthesis of high-entropy polymers, obtain the specific functional properties they are designed for, and so on.

9.6 CONFLICTS OF INTEREST

There are no conflicts to declare.

REFERENCES

[1] Gao M C, Yeh J W, Liaw P K, et al. *High-Entropy Alloys[M]*. Cham: Springer International Publishing, 2016. DOI:10.1007/978-3-319-27013-5.
[2] George E P, Curtin W A, Tasan C C. High entropy alloys: A focused review of mechanical properties and deformation mechanisms[J]. *Acta Materialia*, 2020, 188: 435–474. doi:10.1016/j.actamat.2019.12.015.
[3] Yeh J W, Chen S K, Lin S J, et al. Nanostructured high-entropy alloys with multiple principal elements: novel alloy design concepts and outcomes[J]. *Advanced Engineering Materials*, 2004, 6(5): 299–303.
[4] Tsai M H, Yeh J W. High-entropy alloys: a critical review[J]. *Materials Research Letters*, 2014, 2(3): 107–123. DOI:10.1080/21663831.2014.912690.

[5] Zhang Y, Zuo T T, Tang Z, et al. Microstructures and properties of high-entropy alloys[J]. *Progress in Materials Science*, 2014, 61: 1–93. DOI:10.1016/j.pmatsci.2013.10.001.

[6] Miracle D B, Senkov O N. A critical review of high entropy alloys and related concepts[J]. *Acta Materialia*, 2017, 122: 448–511. DOI:10.1016/j.actamat.2016.08.081.

[7] Zhang W, Liaw P K, Zhang Y. Science and technology in high-entropy alloys[J]. *Science China Materials*, 2018, 61(1): 2–22. DOI:10.1007/s40843-017-9195-8.

[8] Wang X, Guo W, Fu Y. High-entropy alloys: emerging materials for advanced functional applications[J]. *Journal of Materials Chemistry A*, 2021, 9(2): 663–701. DOI:10.1039/D0TA09601F.

[9] Xian X, Zhong Z H, Lin L J, et al. Tailoring strength and ductility of high-entropy CrMnFeCoNi alloy by adding Al[J]. *Rare Metals*, 2022, 41(3): 1015–1021. DOI:10.1007/s12598-018-1161-4.

[10] Zhang W, Li Y, Liaw P k, et al. A strategic design route to find a depleted uranium high-entropy alloy with great strength[J]. *Metals*, 2022, 12(4): 699. DOI:10.3390/met12040699.

[11] Shi Z J, Wang Z B, Wang X D, et al. Effect of thermally induced B2 phase on the corrosion behavior of an Al0.3CoCrFeNi high entropy alloy[J]. *Journal of Alloys and Compounds*, 2022, 903: 163886. DOI:10.1016/j.jallcom.2022.163886.

[12] Li J, Fang Q, Liaw P K. Microstructures and properties of high-entropy materials: modeling, simulation, and experiments[J]. *Advanced Engineering Materials*, 2021, 23(1): 2001044. DOI:10.1002/adem.202001044.

[13] Zhang X, Yao J, Liu B, et al. Three-dimensional high-entropy alloy—polymer composite nanolattices that overcome the strength—recoverability trade-off[J]. *Nano Letters*, 2018, 18(7): 4247–4256. DOI:10.1021/acs.nanolett.8b01241.

[14] Huang Y, Yeh J W, Yang A C M. "High-entropy polymers": A new route of polymer mixing with suppressed phase separation[J]. *Materialia*, 2021, 15: 100978. DOI:10.1016/j.mtla.2020.100978.

[15] Ajitha A R, Thomas S. Chapter 1 — Introduction: Polymer blends, thermodynamics, miscibility, phase separation, and compatibilization[M]. *Elsevier*, 2020: 1–29. DOI:10.1016/B978-0-12-816006-0.00001-3.

[16] Zhang H, Wang Y, Huang J, et al. Low-enthalpy and high-entropy polymer electrolytes for Li-metal battery[J]. *Energy & Environmental Materials*, 2022: e12514. DOI:10.1002/eem2.12514.

[17] Amiri A, Shahbazian-Yassar R. Recent progress of high-entropy materials for energy storage and conversion[J]. *Journal of Materials Chemistry A*, 2021, 9(2): 782–823. DOI:10.1039/D0TA09578H.

[18] Zhang M, Song W, Tang Y, et al. Polymer-based nanofiber—Nanoparticle hybrids and their medical applications[J]. *Polymers*, 2022, 14(2): 351. DOI:10.3390/polym14020351.

[19] Li Z. Applications of polymers in energy conversion and storage fields: A review[J]. *Annals of Chemical Science Research*, 2020, 2. DOI:10.31031/ACSR.2020.02.000535.

[20] Chiu C T, Teng Y J, Dai B H, et al. Novel high-entropy ceramic/carbon composite materials for the decomposition of organic pollutants[J]. *Materials Chemistry and Physics*, 2022, 275: 125274. DOI:10.1016/j.matchemphys.2021.125274.

[21] Fredrickson G H, Liu A J, Bates F S. Entropic corrections to the Flory-Huggins theory of polymer blends: Architectural and conformational effects[J]. *Macromolecules*, 1994, 27(9): 2503–2511. DOI:10.1021/ma00087a019.

[22] Bstes F S, Fredrickson G H. Conformational asymmetry and polymer-polymer thermodynamics[J]. *Macromolecules*, 1994, 27(4): 1065–1067. DOI:10.1021/ma00082a030.

[23] Elmendorp J J. A study on polymer blending microrheology[J]. *Polymer Engineering and Science*, 1986, 26(6): 418–426. DOI:10.1002/pen.760260608.

[24] Ajji A, Utracki L A. Interphase and compatibilization of polymer blends[J]. *Polymer Engineering & Science*, 1996, 36(12): 1574–1585. DOI:10.1002/pen.10554.

[25] Koning C, Van Duin M, Pagnoulle C, et al. Strategies for compatibilization of polymer blends[J]. *Progress in Polymer Science*, 1998, 23(4): 707–757. DOI:10.1016/S0079-6700(97)00054-3.

[26] Wu C F, Arifin D E S, Wang C A, et al. Coalescence and split of high-entropy polymer lamellar cocrystals[J]. *Polymer*, 2018, 138: 188–202. DOI:10.1016/j.polymer.2018.01.064.

[27] Hou X, Chen S, Kohj J, et al. Entropy-driven ultratough blends from brittle polymers[J]. *ACS Macro Letters*, 2021, 10(4): 406–411. DOI:10.1021/acsmacrolett.0c00844.

[28] Hirai T, Yagi K, Nakai K, et al. High-entropy polymer blends utilizing in situ exchange reaction[J]. *Polymer*, 2022, 240: 124483. DOI:10.1016/j.polymer.2021.124483.

[29] Gao L, Song J, Jiao Z, et al. High-entropy alloy (HEA)-Coated nanolattice structures and their mechanical properties[J]. *Advanced Engineering Materials*, 2018, 20(1): 1700625. DOI:10.1002/adem.201700625.

[30] Wang K, Hu J, Chen T, et al. Flexible low-melting point radiation shielding materials: Soft elastomers with GaInSnPbBi high-entropy alloy inclusions[J]. *Macromolecular Materials and Engineering*, 2021, 306(12): 2100457. DOI:10.1002/mame.202100457.

10 Future and Applications

Yong Zhang and Xinfang Song

10.1 INTRODUCTION

High-entropy alloys are a new class of complex materials discovered near the center of the phase diagram and exhibit superior properties compared to conventional alloys. The novel alloy design concept with infinite design space and high-concentration solid-solution structure endow HEAs with excellent mechanical and functional properties, and are expected to achieve unprecedented mechanical and functional properties, making it promising structural and functional materials, such as lightweight alloys, superalloy, and soft magnetic materials. This chapter will discuss future trends and applications of high-entropy materials to improve human life and well-being.

10.2 LIGHTWEIGHT ALLOYS

High-strength and low-density alloys are ideal alloys for engineering applications, especially structural applications, such as aeronautics and civil transportation, where controlling the weight of engineered components is critical to reducing energy requirements [1]. For low-density alloys, it is generally considered to choose low-density Al, Li, Mg, Zn, Sc, Ti, Y, V, Mn, and Cu for combinations. At present, some research results of low-density and high-strength HEAs have been reported. Professor Zhang et al. [2] successfully fabricated $Al_{15}Zr_{40}Ti_{28}Nb_{12}Mo_5$ high-entropy alloys with a density of 6.00 g*cm^{-3} using an arc melting method, which showed excellent compressive mechanical properties, with a compressive yield strength of 1.28 GPa and compressive plasticity of 50%. In addition, a novel precipitation-hardening five-element $Al_{80}Zn_{14}Li_2Mg_2Cu_2$ alloy was prepared by Professor Zhang et al. [3]. Phase structures and properties of alloys are shown in Figure 10.1. The compressive strength of the as-cast $Al_{80}Zn_{14}Li_2Mg_2Cu_2$ alloy is greater than 1 GPa, and the plasticity is greater than 20%. The density of $Al_{80}Zn_{14}Li_2Mg_2Cu_2$ alloy is 3.08 g/cm^3 with synergistic strength and ductility, which breaks the limitations of current lightweight HEAs and high Zn content Al alloys, and expands its application prospects in structural materials.

10.3 SUPERALLOYS

At present, Ni-based alloys are the most widely used alloys in superalloys, with high strength and good oxidation resistance in the range of 650°C~1000°C. The properties of Ni-based superalloys are improved by solid-solution strengthening and γ' phase

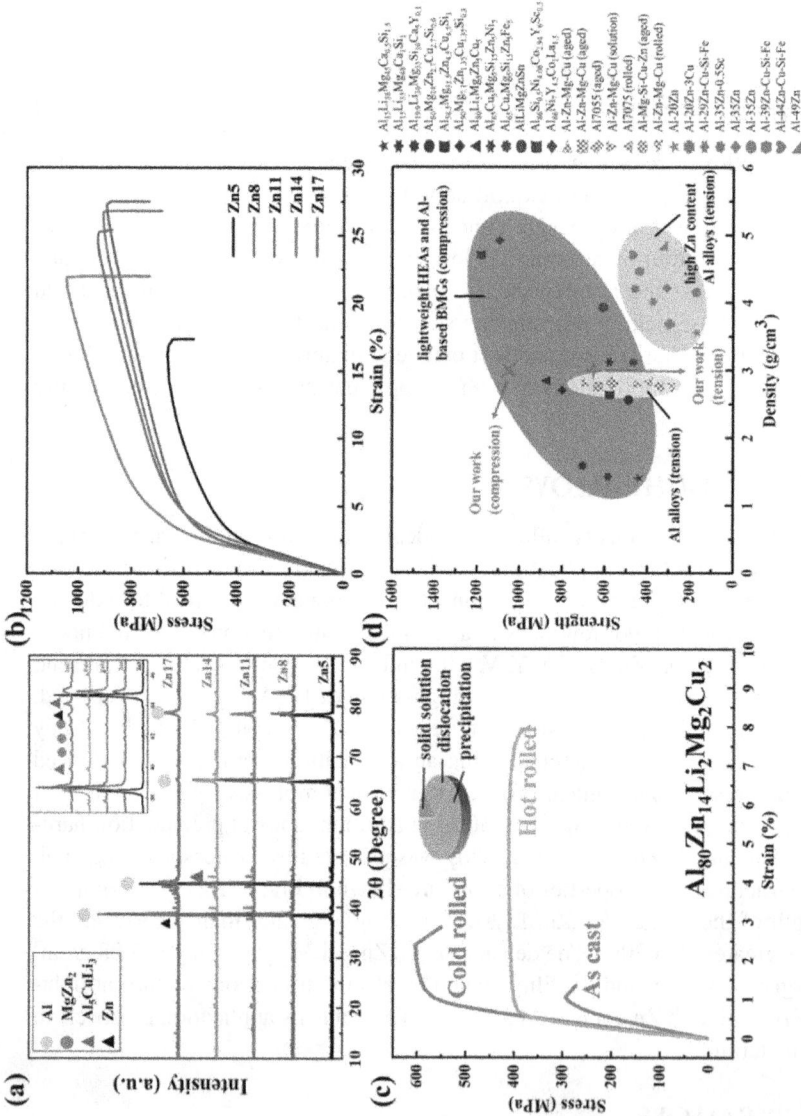

FIGURE 10.1 Phase structures and properties of $Al_{94-x}Zn_xLi_2Mg_2Cu_2$ ($x = 5$, 8, 11, 14, and 17 in at.%) alloys [3]. (a) XRD patterns and (b) the compressive stress-strain curves of as-cast Al94-xZnxLi2Mg2Cu2 alloys. (c) The tensile stress-strain curves of the as-cast, hot-rolled, and cold-rolled Al80Zn14Li2Mg2Cu2 alloy. The inset pie chart describes the main strengthening contributions of the cold-rolled alloy. (d) The comparison of the density and strength between some similar Al-based alloys and the Al80Zn14Li2Mg2Cu2 alloy.

strengthening, of which γ' phase strengthening is the main strengthening method. GH975 alloy is the highest-strength deformed superalloy in the world, but its alloying degree and γ' phase content are close to the design limit.

High-entropy alloys have four major effects and good thermal stability. By adding refractory metal elements, such as W, Hf, Ta, and other high-melting metal elements, novel superalloys with excellent high-temperature resistance can be obtained. Figure 10.2 shows the yield strength-temperature comparison of some high-entropy alloys and conventional superalloys of Inconel 718 and Haynes 230 [1]. It is obvious from the figure that the room-temperature performance and high-temperature performance of Inconel 718 and Haynes 230 are far inferior to those of HEA. Especially, it significantly thermo-softens once the temperature rises above 600°C; in sharp contrast, the refractory HEAs are able to display high strength (> 1 GPa) even at a temperature of 1200°C. In addition, the use of light elements (such as Cr, Ti, and Al) instead of heavy elements (such as Ta and W) to reduce the density of refractory HEA while maintaining excellent high-temperature performance is also one of the future development trends. After the Cr element in the $CrMo_{0.5}NbTa_{0.5}TiZr$ alloy is replaced by the Al element, the density can be significantly reduced by 10.1%, while maintaining a yielding strength up to 2000 MPa at 298 K and 745 MPa at 1273 K, resulting in a very high specific yield strength both at room temperature and at high temperature [4,5]. Therefore, the excellent high-temperature properties exhibited by high-entropy alloys make them a superalloy with great development potential.

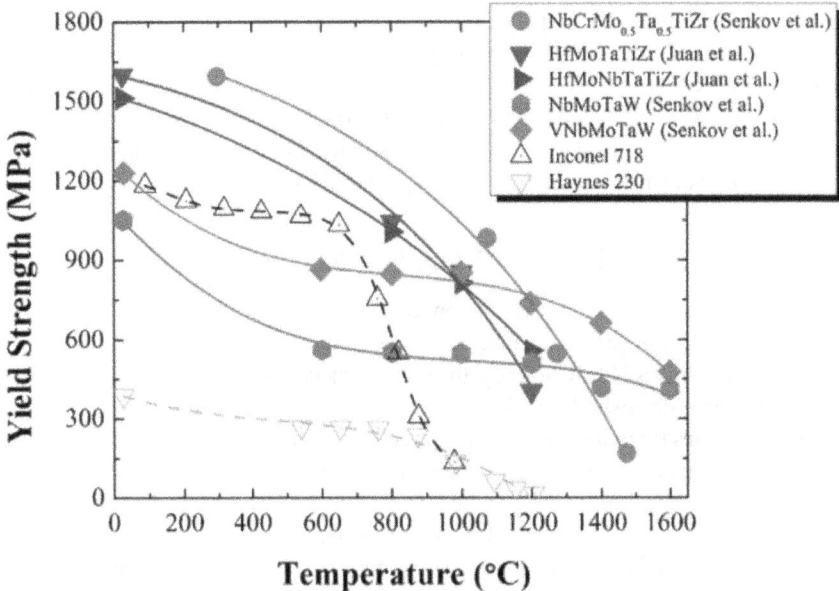

FIGURE 10.2   The yield strength-temperature comparison of some high-entropy alloys and conventional superalloys of Inconel 718 and Haynes 230 [1].

10.4   SOFT MAGNETIC MATERIALS

Traditional soft magnetic materials, such as Fe-Si alloys, have very excellent soft magnetic properties, such as high saturation magnetization, low coercivity, and high permeability [6,7]. However, poor processability and room-temperature brittleness limit
its application, especially in mechanically loaded functional devices [8,9]. Therefore,
it is very important to obtain a soft magnetic material with an excellent combination
of magnetic properties and mechanical properties. Recently, high-entropy soft magnetic materials have begun to develop rapidly. They are mainly composed of three
ferromagnetic elements, Fe, Co, and Ni. And a small amount of one or more
non-ferromagnetic elements are added, such as antiferromagnetic Cr, diamagnetic Cu,
Si, Ga, paramagnetic Al, Ti, Mn, Sn, and other elements. Living up to expectations,
high-entropy soft magnetic materials have been proven to have high strength and ductility as well as excellent soft magnetic properties, and have become one of the most
promising soft magnetic materials. $(Fe_{0.3}Co_{0.5}Ni_{0.2})_{95}(Al_{1/3}Si_{2/3})_5$ alloy is produced by
maglev melting and, subsequently, cold-rolled and annealed at 1000°C, maintaining a
simple face-centered cubic (FCC) solid-solution structure. After annealing, the alloy
exhibits excellent mechanical properties, a tensile yield strength of 235 MPa, an ultimate strength of 572 MPa, an elongation of 38% while maintaining outstanding soft
magnetic properties, a saturation magnetization (Ms) of 1.49 T, and a coercivity of
96 A/m. The emergence of this alloy demonstrates the advantages of soft magnetic
high-entropy alloys in the combination of magnetic and mechanical properties. And
compared with silicon steel and amorphous soft magnetic materials, it also shows the
advantages of easy fabrication and processing and high thermal stability. The comparison of soft magnetic properties between HEAs and traditional soft magnetic materials
can be seen in Figure 10.3 [10]. The multi-component characteristics of high-entropy
alloys can be used to tune the magnetic properties and mechanical properties of soft
magnetic materials so that high-entropy soft magnetic materials have excellent soft magnetic properties and, at the same time, solve the defects of poor mechanical properties of current conventional soft magnetic materials. Industrial applications, such as
motors and transformers, have shown great potential for development.

10.5   RADIATION-RESISTANT MATERIALS

It is well known that particle irradiation could cause atomic displacements, which
induces the irradiation defects, such as vacancies and interstitials, and is also accompa-
nied by thermal spikes [11]. Currently, the phase structure and mechanical properties of
HEAs can remain surprisingly stable under extreme irradiation conditions, making them
popular candidates for radiation-resistant materials, which may be attributed to the effective self-healing mechanisms of HEAs under irradiation conditions, leading to a significantly lower volume swelling rate and defect density of HEAs than that of traditional
alloys [12,13]. Jin et al. [14] studied Ni and Ni-containing FCC-structured hexagonal
alloys and explained the effect of chemical composition on the expansion and hardening
induced by ion irradiation. By controlling the number of elements in the alloy, especially
the element types of alloy, the radiation resistance at 500 °C is improved. Compared with
alloys containing Co and Cr, alloys containing Fe and Mn elements have a greater effect
on the reduction of expansion ratio, as shown in Figure 10.4. By comparison, the quinary

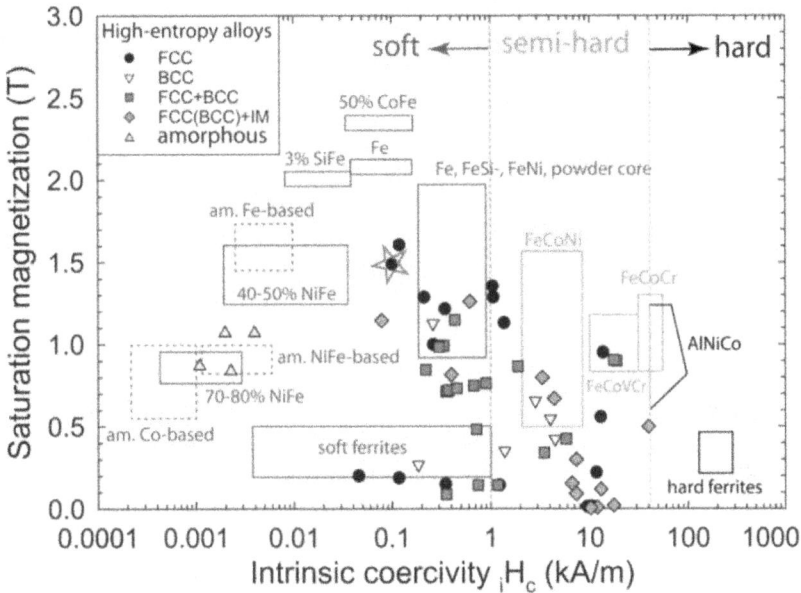

FIGURE 10.3 The saturation magnetization and intrinsic coercivity of the $(Fe_{0.3}Co_{0.5}Ni_{0.2})_{95}$ $(Al_{1/3}Si_{2/3})_5$ alloy compared with other traditional soft magnetic materials [10].

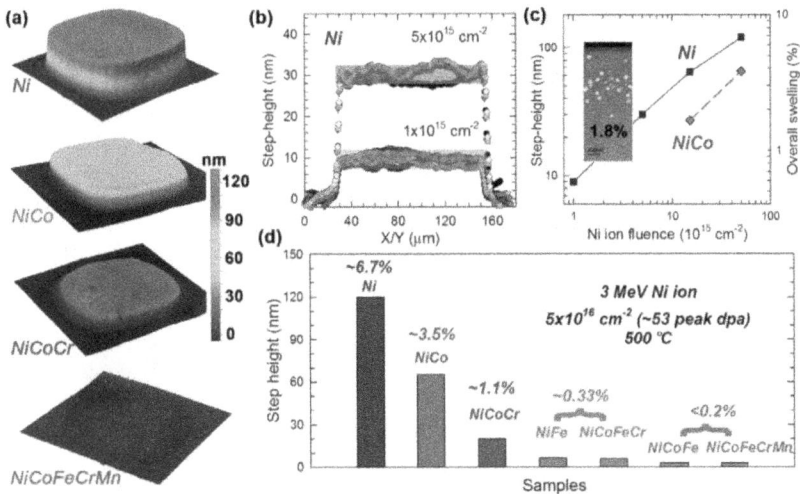

FIGURE 10.4 (a) Surface step measurements of Ni, NiCo, NiCoCr, and NiCoFeCrMn after ion irradiations. (b) Step-height profiles for Ni under low-fluence irradiations. The spots are the data, and the bands indicate the averages and uncertainties. (c) Overall swelling from step-height as a function of ion fluence in Ni and NiCo; the inset is the cross-sectional transmission electron microscopy image for Ni irradiated at a fluence of 5×1015 cm^{-2}. (d) Comparison of step-height and overall swelling among the seven materials [14].

alloy NiCoFeCrMn has good mechanical properties, and its radiation resistance is 40 times that of Ni. Therefore, high-entropy alloys have excellent performance in radiation resistance. By carrying out the multi-principal alloying design of existing radiation-resistant alloys, the radiation resistance properties of alloys can be further improved, which is of great significance for future scientific research and technical applications.

In conclusion, the unique design concept of high-entropy alloys endows it with outstanding structural properties and excellent properties, but research in high-entropy alloys is still in its infancy. There is still a lot of research space, and it will show a broader application prospect in future research.

REFERENCES

[1] Ye Y F, Wang Q, Lu J, et al. High-entropy alloy: challenges and prospects[J]. *Materials Today*, 2016, 19(6): 349–362.

[2] Li Yasong, Liaw Peter K, Zhang Yong. Microstructures and properties of the low-density $Al_{15}Zr_{40}Ti_{28}Nb_{12}M(Cr, Mo, Si)_5$ high-entropy alloys[J]. *Metals*, 2022, 12: 496.

[3] Li R, Ren Z, Wu Y, et al. Mechanical behaviors and precipitation transformation of the lightweight high-Zn-content Al-Zn-Li-Mg-Cu alloy[J]. *Materials Science and Engineering: A*, 2021: 802.

[4] Senkov O, Senkova S, Woodward C. Effect of aluminum on the microstructure and properties of two refractory high-entropy alloys[J]. *Acta Materialia*, 2014, 68: 214–228.

[5] Senkov O, Woodward C, Miracle D. Microstructure and properties of aluminum-containing refractory high-entropy alloys[J]. *JOM*, 2014, 66(10): 2030–2042.

[6] Sankara Subramanian A T, Meenalochini P, Suba Bala Sathiya S, et al. A review on selection of soft magnetic materials for industrial drives[J]. *Materials Today: Proceedings*, 2021, 45: 1591–1596.

[7] Talaat A, Suraj M V, Byerly K, et al. Review on soft magnetic metal and inorganic oxide nanocomposites for power applications[J]. *Journal of Alloys and Compounds*, 2021: 870.

[8] Babuska T F, Wilson M A, Johnson K L, et al. Achieving high strength and ductility in traditionally brittle soft magnetic intermetallics via additive manufacturing[J]. *Acta Materialia*, 2019, 180: 149–157.

[9] Nartu M S K K Y, Jagetia A, Chaudhary V, et al. Magnetic and mechanical properties of an additively manufactured equiatomic CoFeNi complex concentrated alloy[J]. *Scripta Materialia*, 2020, 187: 30–36.

[10] Zhang Y, Zhang M, Li D, et al. Compositional design of soft magnetic high entropy alloys by minimizing magnetostriction coefficient in $(Fe_{0.3}Co_{0.5}Ni_{0.2})_{100-x}(Al_{1/3}Si_{2/3})_x$ system[J]. *Metals*, 2019, 9(3).

[11] Xia S Q, Yang X, Yang T F, et al. Irradiation resistance in AlxCoCrFeNi high entropy alloys[J]. *JOM*, 2015, 67(10): 2340–2344.

[12] Jin K, Lu C, Wang L, et al. Effects of compositional complexity on the ion-irradiation induced swelling and hardening in Ni-containing equiatomic alloys[J]. *Scripta Materialia*, 2016, 119: 65–70.

[13] Nagase T, Rack P D, Noh J H, et al. In-situ TEM observation of structural changes in nano-crystalline CoCrCuFeNi multicomponent high-entropy alloy (HEA) under fast electron irradiation by high voltage electron microscopy (HVEM)[J]. *Intermetallics*, 2015, 59: 32–42.

[14] Jin K, Lu C, Wang L M, et al. Effects of compositional complexity on the ion-irradiation induced swelling and hardening in Ni-containing equiatomic alloys[J]. *Scripta Materialia*, 2016, 119: 65–70.

Index

For Product Safety Concerns and Information please contact our EU
representative GPSR@taylorandfrancis.com
Taylor & Francis Verlag GmbH, Kaufingerstraße 24, 80331 München, Germany

www.ingramcontent.com/pod-product-compliance
Lightning Source LLC
Chambersburg PA
CBHW070728220326
41598CB00024BA/3355